Pardons

PARDONS

Justice, Mercy, and the Public Interest

KATHLEEN DEAN MOORE

New York   Oxford
OXFORD UNIVERSITY PRESS
1989

Oxford University Press

Oxford   New York   Toronto
Delhi   Bombay   Calcutta   Madras   Karachi
Petaling Jaya   Singapore   Hong Kong   Tokyo
Nairobi   Dar es Salaam   Cape Town
Melbourne   Auckland

and associated companies in
Berlin   Ibadan

Library of Congress Cataloging-in-Publication Data
Moore, Kathleen Dean.
Pardons: justice, mercy, and the public interest/ Kathleen Dean Moore.
p.   cm.   Bibliography: p.   ISBN 0-19-505871-2
1. Pardon—United States.   2. Pardon.   I. Title.
KF9695M66   1989 345.73'077—dc19   [347.30577]
88–30286   CIP

Excerpts from *Doing Justice* by Andrew von Hirsch,
copyright © 1976 by Andrew von Hirsch, are reprinted by permission
of Hill and Wang, a division of Farrar, Straus and Giroux, Inc.

9 8 7 6 5 4 3 2 1

Printed in the United States of America
on acid-free paper

*I dedicate this book with love
to my father and teacher,
Donald S. Dean,
a man of integrity and compassion*

Acknowledgments

This is a book about the philosophical justification for the practice of pardoning people who have broken the law. My central theme is that acts of pardon need to be justified, and that their justification is to be found in the very principles of retributive justice which make it reasonable to punish in the first place. My hope is that this book will stimulate further philosophical work that will ultimately earn pardon the kind of philosophical scrutiny accorded to punishment, and that pardoning practices will become more principled as a result.

I wish to acknowledge the assistance of several institutions which aided in the completion of this volume. The National Endowment for the Humanities, through a Travel-to-Collections Grant, funded my trip to Washington, D.C., where I studied the materials in the files at the Office of the Pardon Attorney in the Department of Justice. The Pardon Attorney's staff was particularly helpful and gracious, making their files and work space available to me. Harvard Law School allowed me to use its library, an inexhaustible resource and an inspiring place in which to write. The Philosophy Department at Oregon State University provided materials and supplies, secretarial assistance, and some travel funds.

The person who deserves most credit for what is good about this book is my husband, Dr. Frank L. Moore, a scientist. He read every draft of the manuscript; many of the book's arguments are sounder, many sentences clearer, and many bad jokes missing, as a result of his good sense and high standards. More important, he supplied the encouragement, confidence, free time, and strategic advice that are necessary to see a long and sometimes discouraging project through to completion.

I am grateful to our children, Erin and Jonathan, who lived with the fact that every hour spent on this book was an hour spent away from them and who never lost faith in me or in my work. My father, to whom this book is dedicated, helped in ways both tangible and intangible. By his constant monitoring of the media for news about pardons and politics, he kept me informed; by his unconditional pride in me, he kept me going.

The academic tradition whereby a perfect stranger agrees to spend hours carefully examining someone else's arguments for no tangible reward and at great cost in time and effort is a wonder to me. Professor Michael Davis, a philosopher at the Illinois Institute of Technology, was one such stranger. He provided a thorough and insightful critique of the entire manuscript. Professor Claudia Card, a philosopher at the University of Wisconsin, is another who read the entire manuscript and offered suggestions which proved to be enormously helpful. Professor Albert Alschuler, my criminal law professor now teaching at the University of Chicago School of Law, critiqued Part I, saving me from some real blunders and directing me to new sources of information.

My colleagues at Oregon State University—Professors Robert Dale, Jon Dorbolo, Flora Leibowitz, Peter List, Michael Scanlan, and William Uzgalis—all read parts of early drafts of my manuscript and provided criticism and encouragement in equal measure. Professor William Sacksteder, my dissertation advisor at the University of Colorado, helped shape the arguments and polish the prose in some parts of the book having to do with retributivist theory.

For what is traditionally called 'technical assistance,' I wish to thank Janie Hamblin, who photocopied endless numbers of pages, who typed early drafts of the manuscript, and who hopes that, next time, I will do her the favor of writing a romance novel.

Corvallis, Oreg.   K. D. M.
October 1988

Contents

III   Applications, Both Practical and Theoretical

Pardons

Introduction

> To all to whom these presents shall come, greeting: . . .
>
> *Whereas* it does not appear that the ends of justice require that the [offender] serve the aforesaid sentence in its entirety;
>
> NOW, THEREFORE, BE IT KNOWN, that I, . . . , the President of the United States, in consideration of the premises, divers other good and sufficient reasons me hereunto moving, do hereby commute the aforesaid sentence. . . . [1]

These are welcome words to a convicted felon, for they begin an Executive Grant of Clemency. Presidential and gubernatorial acts of clemency—grants of pardon, commutations, and reprieves—are granted for a wide variety of reasons, some good and some not. These four cases provide a sample:

In celebration of Thanksgiving Day, 1912, South Carolina Governor Coleman Blease granted executive clemency to a convicted murderer named William H. Mills. The Governor was fulfilling a promise made to his constituency while campaigning for election. If you elect me, he had told a rowdy group of supporters, I'll free anybody you want. The crowd wanted Mills. Mills it is, the Governor had said, but make sure I get elected. The Governor won the vote; in due course, Mills won his freedom. [2]

Robert Q. Acker, alias Robert I. Acker, was convicted of robbing the U.S. mail. He served his full sentence of eighteen months in the Kalispell County, Montana, jail. It appeared that he had been "continuously employed since his release and otherwise conducting himself as a law-abiding citizen. The Attorney General advised [the President] that a full and unconditional pardon be granted for the purpose of restoring his civil rights." [3]

Vernon Atchley killed his wife and was sentenced to die in 1962 in a California electric chair. Governor Edmund G. Brown refused to commute the sentence; it had been a bloody and senseless murder, and even governors opposed in principle to the death penalty are able to feel revulsion and to sense the direction of political winds. Governor Brown did, however, order a last-minute electroencephalogram. The test showed significant brain damage, probably the result of a barroom brawl. The Governor changed his mind, commuted the sentence, and ordered EEGs for all death row prisoners.[4]

Explaining that he was responding to his conscience, President Gerald Ford fully pardoned forever all of Richard Nixon's crimes, known or hidden. He acted partly from fear that the ex-President would not receive a fair trial and partly from compassion for the disgraced man's family and the man himself, who had already suffered an unprecedented loss. Primarily, however, President Ford wanted to end the divisive discussion, to get on with more pressing business.[5]

These examples illustrate just a few of the many different kinds of reasons behind acts of clemency. It is clear that some of those reasons are better than others. Granting a pardon on a whim to please a crowd is a corruption of the clemency power; on the other hand, relieving the punishment of a person too incapacitated to be blamed is a praiseworthy act. Whether Richard Nixon's pardon was wise is debated to this day.

What is not clear is the basis on which judgments about the moral justifiability of pardons are to be made. In this book, I offer a theory of pardons that shows how the principles of retributive justice can be used to distinguish between cases in which a pardon is justified and even morally obligatory, and cases in which a pardon is an abuse of the clemency power.

What Are Acts of Clemency?

In its broadest sense, an act of clemency takes place whenever someone gives up a moral or legal claim against another. But clemency can be understood in a more restricted sense. An act of clemency, as it constitutes the subject of this book, is an official act by an executive that removes all or some of the actual or possible punitive consequences of a criminal conviction. It ensures that the person on whom it is bestowed will not suffer all the punishment the law usually inflicts for the crime committed.[6]

An executive who wants to exercise clemency has several options. One option, the *pardon*, can take several forms. It can be a full pardon; fully pardoned offenders simply walk away from jail as if they had never been tried and sentenced. It can be partial, relieving the offender of some but not all of the legal consequences of conviction. It can be absolute, freeing the criminal without any conditions whatever. Or it can be conditional, dependent on the performance or nonperformance of acts specified by the executive. It can be individual—pardoning a particular person—or general—pardoning a class of people.

Usually a pardon follows trial, conviction, and sentencing. But one may be pardoned before he or she has been formally accused of a crime, as President Ford taught the nation.

Executives can also grant *amnesty*, a declaration that they are going to "forget about the whole thing." Amnesties often can be distinguished from pardons by the fact that amnesties are usually granted to groups of people, unnamed members of a class defined by a particular sort of offense they all committed (desertion, usually, or draft evasion). But the practice of general pardons blurs this distinction. Usually amnesties are granted before a trial takes place; this is sometimes taken to be a distinguishing characteristic. But since pardons can precede conviction, this also is a distinction without a difference.

Reprieves, often categorized with pardons as acts of clemency, are a less valuable gift. The reprieve postpones execution of the sentence for a specified period of time—until a pregnant death-row inmate's child can be born, for example, or until a sick capital offender gets well enough for execution, or until all appeals are heard. But a reprieve does not prevent the punishment altogether.

A *commutation* substitutes a lesser for a more severe sentence. Thus, punishment takes place, but in a reduced form. Presidents occasionally commute a sentence to the time already served or a death penalty to a life sentence.

All these are acts of clemency—official use of the executive power to reduce the severity of a punishment at the discretion of the executive. They differ only in degree or procedure. These acts are the subject matter of this book. For convenience, I use the word 'pardon' to refer to all these forms of official clemency.

In the United States, the President holds the power to pardon federal offenses. In actual fact, most pardon applications are processed by the Office of the Pardon Attorney in the Department of Justice. Warrants of pardon for the chosen few are presented in large batches to the President, who often signs without even reading them. There have been, of course, occasions when Presidents themselves have initiated pardons in special cases.

In most states, the pardon power belongs to the Governor, who often shares the power in some way with a Board of Pardons. The central focus of this book is on presidential pardoning power, taken to be representative of other exercises of clemency. Nevertheless, this focus does not prevent occasional references to states and to governors, and indeed these provide some of the most garish examples of what it means to abuse pardons.

I should make clear at the start that parole is not part of my subject matter. It is true that pardon and parole are often linked together in legal publications, and some states have combined pardon and parole functions in one decision-making body. It is also true that parole and some kinds of conditional pardons have the same practical effect.

Nevertheless, to link them conceptually is to invite confusion. Parole is a kind of punishment, which decreases in severity as it comes to a close. "A parolee is a convicted criminal who has been sentenced to a term of imprisonment and who has been allowed to serve a portion of that term outside prison walls."[7] As a part of punishment, parole practices are highly rule-bound and subject to judicial review and constitutional protections. Judges, prison officials, and parole boards make decisions about paroles. In contrast, presidents and governors make decisions about pardons. Moreover, because pardons relieve punishment, pardon practices have traditionally not been limited by rules or judicial review; nor have those offenders denied pardons been able to complain that their rights have been violated. It may be that good reasons for granting parole are the same as good reasons for granting pardon, but this cannot be presupposed. Therefore, it is advisable to confine this book's discussion to pardons.

The Need for Philosophical Discussion of Pardons

It is surprising that pardons have not received wider scholarly attention. Pardons were an important topic for most of the great Enlightenment philosophers. But until very recently, pardons have proceeded in relative philosophical obscurity in the nineteenth and twentieth centuries. There is a pressing need to make up for lost time.

Pardons are more important than is often recognized. Because they are usually granted quietly, pardons occur with a far greater frequency than most people would ever expect. They exert some influence on the course of human events, an influence that can only increase as anticipated changes in U.S. criminal law take place. Not least importantly, pardoning practices raise fundamental questions in the philosophy of law.

Pardoning is a practice common to all cultures and all periods of history. With the exception of China's, all the constitutions in the world provide for a pardoning power.[8] Pardons take place in forty-eight of the fifty United States, literally by the hundreds each year. U.S. Presidents have granted an average of 104 acts of clemency per year over the past ten years,[9] a statistic that comes as a surprise to many people.

Pardons have a profound effect on the lives of the people who are pardoned, of course, but their influence extends beyond that. The papal abuse of indulgences—the high price sinners were required to pay to the church for divine forgiveness—was one of the factors that led to the Protestant Reformation and the subsequent political upheaval in Europe. President Ford's pardon of Richard Nixon may have cost him reelection; several Governors have been impeached or driven from office for abusing their power to pardon.

It is reasonable to expect that the role of pardons will expand in the next decade because of changes now taking place in the criminal law. These changes grew out of widespread disillusionment with the failures of a penal system based on reformist ideas—a system in which judges had the power to assign individualized, indeterminate sentences in 'correctional facilities,' until offenders could safely and gradually be reintroduced into society by means of furloughs and parole. Under the new rules, judicial discretion will be sharply curtailed by sentencing guidelines, and parole will be gradually eliminated. Pardon may end up as the only means of individualizing sentences.

Pardoning practices also deserve more philosophical attention for a different reason: They are fertile ground for raising perennially important philosophical issues. The four cases cited at the beginning of this chapter are thick with philosophical issues. For example: How can a single act be both kind and immoral? What makes a person a 'new man' with a clean slate? Why should feeble-mindedness keep someone from being executed, if it did not keep him from murdering his wife? What does a person deserve who does harm by trying to do what he thinks is right? What difference does it make that someone has already suffered for his offense? These questions lead into ongoing debates about the justification for punishment, the nature of justice, and grounds for responsibility.

I am drawn to this subject for another reason as well—its melodrama. The major roles in pardoning are played by constitutional scholars and mass murderers, ambitious politicians and conscientious objectors, Washington bureaucrats, pathological liars, pregnant felons, and philosophy professors. The drama takes place in newspaper headlines, prison cells, meetings of powerful people. For once, it is no exaggeration to say that here is a philosophical issue that is a matter of life and death. Why, then, have pardons received relatively little scholarly attention? There are several possible explanations:

First, the presidential pardoning power is not subject to review by the courts, so the subject does not often come up for discussion by judges. Nor can a pardon be overturned by Congress, by the people, or even by the President himself. Perhaps scholars turn away from the subject of pardons thinking their ideas will make no difference. It seems to me that this is misguided. That pardons cannot be reviewed by courts is a good reason why they *should* be reviewed by scholars; a free-ranging debate about the proper uses of a presidential power is useful just because the power is absolute.

Second, presidential pardons proceed quietly for the most part (with obvious exceptions). The interesting details of pardon applications are concealed to protect the privacy rights of the applicants and perhaps to diminish the scrutiny the pardons might otherwise receive. As a matter of fact, the large majority of pardons are routine. Although they raise fascinating philosophical and policy questions, they make poor press; in the typical case, a felon serves time, leaves jail, leads an exemplary life for five years, and then applies for a pardon, hoping that the action will help remove some of the civil disabilities resulting from a felony conviction, or that he will be redeemed in his children's eyes before he dies. Pardons like this attract little attention. But again, the very quietness of the pardoning process is a reason for more, rather than less, attention.

I suggest, though, that the overriding reason why there has not been adequate philosophical discussion of pardons is a philosophical one: Pardon has historically been understood as an act of grace, a gift freely given from a God-like monarch to a subject. Gift-giving is not something to criticize, analyze, scrutinize, demand, refuse, or justify. When a child gives a parent a gift, saying, "I made it for you because I love you," that ends any discussion of whether it is a good gift or not. If pardoning is likewise a supererogatory act, that may be the last important thing that needs to be said about it.

Herein lies the most important reason why pardons should receive greater philosophical attention. It is my view that pardons are duties of justice, not supererogatory acts. If pardoning is *not* a species of gift-giving—as I believe it is not—then a great deal of work remains to be done, in order to explain and justify the practices of pardon.

An Adequate Theory of Pardon

The people who make the daily decisions about who should be pardoned for federal offenses, and who not, clearly believe that a pardon should be understood as a free gift.[10] This gift-giving theory of pardon has descended, conceptual baggage intact, from the God-like powers of an absolute monarch. There is serious question about whether that view is adequate. For an adequate theory of pardon should do at least the following:

1. It should allow room for criticism of pardons; it should specify the grounds for criticism of pardons; and it should make clear why those grounds for criticism count. For it must be admitted that, unlike children's gifts, some pardons are wrong, unjust. It is possible to abuse the pardoning power, as Governor Blease surely did in South Carolina and as some would argue President Ford did. A theory of pardon that does not acknowledge this possibility is not doing its job.
2. An adequate theory of pardon should clarify the relationship between the justification for punishing and the justification for pardoning.
3. An adequate theory of pardon should be grounded in a general moral view and in a political philosophy—a systematic view, that is, of what is right to do and of what is right *for the state* to do.
4. An adequate theory of pardon should have practical application, providing a source of suggestions for ways the pardoning practice could be improved.

The gift-giving concept of pardon can do none of these things. Accordingly, this essay is an attempt to work out a new theory of pardons, based on the ideas of retributive justice. The essay addresses a central question: Given a retributivist theory of justice and of the role of the state, under what conditions is a pardon justified, and under what conditions is a pardon wrong?

The Plan of the Book

This book consists of three parts. The first part provides the context—philosophical, constitutional, and historical—for the ensuing discussion. The context is rich; this work steps into a dialogue that began long before the classical Greeks, continued angrily during the time when pardons could be bought from James II for two shillings and from God for considerably more, and quietly continues in the American courts and classrooms of the twentieth century.

The story of pardon and retribution is a complex one, since it is necessary to follow three concurrent plot lines. First, there is the history of the pardoning power itself, a story of powerful people using their power. In addition, there is the evolution of the concept of pardon through Supreme Court decisions. And finally, there is an ongoing devising and revising of theories of punishment and, occasionally, theories of pardon by philosophers and legal scholars. What makes the story particularly fascinating is the interplay between theory and practice, as they chase through the centuries, first one and then the other taking the lead.

Part I begins by describing a practice—the great monarchs of Europe shamelessly abusing 'acts of grace' to enrich themselves and populate their colonies—that was excoriated by the retributive theories of Immanuel Kant and others. It follows the changing rationales for punishment and pardon through the intervening centuries, explaining how these relate to the ways pardons were used. The story ends in the present, explaining how a sadder-but-wiser retributivism points the way toward a justification of pardon based on what an offender deserves from the state, at the same time as retributivist-inspired changes in sentencing procedures make pardons essential.

Part II is an attempt to work out the details of a retributivist justification for pardon. Any retributivist justification of anything is based on the principle that a person should get what he deserves.

This turns out to be a difficult formula, for retributivists do not agree on appropriate grounds for desert. On the one hand, 'legalistic' retributivists argue that what people deserve from the state is based on what they have done. In contrast, 'moralistic' retributivists argue that what people deserve from the state depends on what kind of people they are, a question of moral desert. Combining both theories, I suggest that pardons are required when a convicted person is not liable to punishment on retributivist grounds; that pardons are justifiable when a convicted person is liable to punishment, but not morally deserving of punishment; and that pardons are unjustifiable when a convicted person is both liable to punishment and morally deserving.

Working out the practical consequences of the theoretical formula turns out to be enormously difficult. I suggest that pardons may be appropriate on retributivist grounds under these conditions:

1. Innocence: (a) to override a false conviction, or to prevent the punishment of a person whose guilt is in substantial doubt; (b) to prevent the punishment of an 'innocent,' a person who cannot be blamed because of substantially reduced ability; or (c) to remove the stigma of a felony conviction from a person who has, over time, become a 'new person.'
2. Excusable crime: when the offender gained nothing from the crime, either because (a) he acted unintentionally and made full reparations; (b) he was the only victim of his crime; (c) his crime repaired rather than created an injustice; or (d) the crime was coerced.
3. Justified crime: (a) to reduce punishment for criminal acts conscientiously performed; and (b) to prevent the punishment of morally justified acts.
4. Adjustments to sentences: (a) to relieve the punishment of an offender who has suffered enough, or one whose particular circumstances would make him suffer more than he deserves; or (b) to prevent an unwarranted cruel punishment.

Part III explores some of the conclusions that can be drawn from the theory of pardons—conclusions both theoretical and practical. The first conclusions concern the ongoing philosophical dispute about how forgiveness, mercy, and pardon can be distinguished; I suggest that the attitude of forgiveness can be distinguished from mercy and pardon, which are acts. Mercy is a private act of relinquishing a legitimate claim against another for motivations having to do with pity or generosity. Pardon alone is an official act, a performative utterance, which has the effect of relieving the consequences of criminal conviction.

If all this is taken seriously, it becomes clear that changes ought to be made—changes in the way people think about pardons and changes in the way the President uses the pardoning power. Pardons can no longer be thought of as free gifts; nor are pardons discretionary. Granting a pardon is a duty of justice that follows from the principle that punishment should not exceed what is deserved. Moreover, it should not automatically be assumed that a pardon implies guilt; in fact, a pardon is more likely to imply a degree of innocence. Pardons should only be granted after sentencing, and pardons should not be valid until they are accepted by the person pardoned. All these, I argue, are consequences of taking the view that pardoning should be consistent with retributivist principles of justice.

The presidential pardoning power needs close scrutiny—not the sort of scrutiny of the Gallup Poll, but a theory-based scrutiny that both clarifies when the power is abused and explains why. It is characteristic of pardons that they occur in clusters, usually right after disasters of one sort or another. Again and again—after the Civil War, after Prohibition, after the Vietnam War, after Watergate, perhaps even after the Iran-*contra* arms deal—controversies about pardons arise when people are least able to think dispassionately about them. Ungrounded in principle, their arguments tend to be *ad hoc* and angry. This book is an effort to sort the arguments out, to understand pardon in the context of a theory of justice, in the calm before the next crisis.

I

A Philosophical History
of Punishment and Pardon

1

Pardon Before the Enlightenment

As long as people have been thinking about punishment, they have been thinking about the remission of punishment. Western history gives no reason to think that the urge to punish is any older or any stronger than the impulse to pardon. Provisions for pardon were present in even the earliest legal codes. And it is arguable that pardons have played as pivotal a role in history as their better-known opposite, punishments.

Pardon in Ancient Legal Codes

Retributive justice and pardon were bound together in the oldest known legal code—the Code of Hammurabi, developed by the Babylonians around the eighteenth century B.C. The Code is most famous for its startling list of punishments: mutilation or amputation of offending body parts, enslavement, drowning, and so on. But the Code is more remarkable for its efforts to put a limit on private revenge, blood feuds, and even official punishment. Accidents were not to be punished; people guilty of manslaughter, for example, were merely fined. Certain excuses and justifications were allowed; deserted, poverty-stricken wives, for example, were not punished for bigamy. Article 129 specified circumstances under which adulterers were pardoned.[1]

Most important, all sentences were limited by the principle of *lex talionis*, the law of retaliation. "A life for a life," the Code decreed, but nothing more. The Code was a serious attempt to make punishment proportional to the perceived severity of the crime.[2] If a nobleman broke the bone of a nobleman, then indeed the offender's bone was to be broken. But were the nobleman foolish enough to strike someone of even higher rank, his penalty was sixty lashes with an oxtail whip. On the other hand, if a slave struck the nobleman, the slave lost an ear. Such were the delicate gradations of harm inflicted and received.

The Old Testament, often viewed as the story of a vengeful God, can be viewed also as a chronicle of mercy. Indeed, the oldest crime in history, Cain's murder of his brother, was not fully punished; God commuted Cain's banishment by allowing him to live in the land of Nod.[3] "Merciful the Lord is, and just, and full of pity."[4] Pardons granted (and sometimes treacherously revoked) by the Kings of Israel and the limited amnesty afforded those who could reach a place of safety were evidently well-established traditions in Old Testament times.

Comparatively little is known about pardons in ancient Athens; ironically, there is no Greek word for 'pardon,' even though the word 'amnesty' is a gift from the Greeks. The great Athenian philosophers had little to say about pardons.[5] In Athens, the power of pardon rested with the people. To obtain a pardon, a supplicant had to have the signatures of 6,000 citizens. It will not surprise anyone today to learn that only in the cases of famous athletes, popular actors, and disgraced rulers could that many interested people be found.[6]

The Romans had a sophisticated and frequently used system of pardons—legislative and judicial, full and partial, pardon by the will of the Emperor or by accidental encounter with a vestal virgin.[7] A peculiar form of pardon was used to soften the punishment of mutinous companies of soldiers. Unwilling to execute an entire army, a commander would sometimes order his men to count off in tens, pardon the first nine and kill each tenth soldier. Thus were mutinous troops 'decimated.'

Perhaps the most notorious Roman pardon power was that exercised by Pontius Pilate. "But you have a custom that I should release unto you one at the Passover,"[8] he said to the crowd at the Passover feast. The crowd, of course, chose a murderer and rebel, Barabbas, and thus declined the opportunity to change the plot of the New Testament. Like the Jews, the Romans were accustomed to pardoning (and executing) criminals on coronation days and local holidays, when the colonial rulers were most interested in subduing crowds. The Romans evidently understood that the power to pardon is every bit as great a power as the power to punish, and they used the pardon often and skillfully for their political ends.

Pardoning Power in Europe and England
Before the Enlightenment

From the Fall of the Roman Empire to the Enlightenment in the eighteenth century, pardoning practices in Europe and England were varied and complicated. So rich is the history of the pardoning power during this time, and so important was the pardoning power to this period of history, that a full accounting cannot be attempted here. Certain general observations can, however, be made.

Throughout this period, the power to pardon was recognized as a great power indeed; when rival authorities struggled for power, it was often the power to pardon that they sought. Everyone who claimed a right to punish—and there were many, from Druid priests to archbishops, from clan chiefs to kings, from mobs to legislatures and courts—also claimed a right to pardon.

In England, for example, the prerogative of mercy was assigned by law to the King as early as the seventh century, and it was still the King who held the pardoning power under the new code of laws of William the Conqueror. But the power did not go uncontested. The Earls held a rival pardoning power in their own lands, and the church also had the right to pardon offenses through the 'benefit of clergy.' The Parliament also regularly contested the King's pardoning power.[9] Thus in England, as in much of Europe, the unceasing struggles for power often took the form of struggles for the exclusive power to pardon, which was correctly recognized as of great value, politically and financially.

The pardoning power had an essential role to play in justice systems in which the criminal law was severe by any standard. In England, death was the penalty prescribed for every felony; by 1819, fully 220 offenses were capital offenses. The liberal use of the royal prerogative of mercy softened the harshness of the system. So, as Albert Alschuler pointed out,[10] of the 1,254 defendants sentenced to death in England in 1818, only ninety-seven were executed. The others received the King's pardon, often on the recommendation of the trial judge.

Moreover, the pardoning power gave effect to distinctions that were either lacking or insufficiently developed in the young system of criminal law—distinctions between manslaughter and murder, for example, or between intentional and accidental harms. Excuses (such as youth, mistake, insanity) or justifications (such as necessity and self-defense) were recognized grounds for pardons, long before they became grounds for acquittal.

The fate of two men in Staffordshire, England, in 1293 illustrates the importance of the pardoning power to the whole scheme of justice: In the dark of night, two men were chasing a thief. Each mistaking the other for the culprit, they tackled each other and began to fight. One was killed. The hapless 'murderer' was sentenced to death and given friendly advice to seek a pardon from the King.[11]

Likewise, when a child over the age of seven killed someone, he had nowhere to turn for mercy but to the King. In 1748, a ten-year-old boy murdered a five-year-old and buried his body in a dung heap. The verdict: guilty. The sentence: death. The small murderer won several reprieves and eventually a full pardon, the King being able to do what the strict criminal law could not.

In France as well, all homicides were murders and all murders were punishable by death. In her fascinating account of pardon in sixteenth-century France,[12] Natalie Davis explained both the elaborate process by which supplicants appealed for the King's mercy and the extensive guidelines that determined whether or not it would be granted. A supplicant might be pardoned if his homicide was justified, as for example, if he was avenging the adultery of his wife. Or he could be pardoned under any of eleven other circumstances, including "homicide during a tennis game" or "homicide committed by a person who is rare and excellent."[13] Thus the pardons served useful purposes.

However, supplicants were not the only ones who benefited from pardons. The person granting the pardon stood to gain some advantages, not the least of which were financial. These were the days when pardons were granted freely, but almost never for free. Despite the fact that the Magna Carta forbade sales of justice, pardons became a dependable source of income for the church and state. There was apparently no disgrace in offering or accepting money for a pardon. The officers of the church did it, selling papal indulgences to finance their petty wars and not-so-petty lifestyles.[14] The Kings did it, selling pardons for prices ranging from two shillings to, in one case, 16,000 pounds sterling, of which James II got "one half and the other half was divided among the two ladies then most in favor."[15] One Mayor received "one hundred tuns of wine" in exchange for a pardon.[16] In France, the cost of a letter of remission was equivalent to two months' wages for an unskilled laborer.[17]

By the seventeenth century, the conditional pardon was put to another use: easing the chronic labor shortages in the colonies. Capital offenders were sometimes offered a pardon, on condition that they agree to go to the New World and work on the plantations for a number of years.[18] Thus, the conditional pardon was the origin of a new kind of punishment—transportation—a practice which no doubt explains why so many of us were born in America. It is difficult to determine how many felons were sent to the New World; estimates vary between 15,000 and 100,000. By 1663, so many felons lived in Virginia that there was constant danger of rebellion.[19] Also, the ships of the growing British navy were staffed in part by reluctant sailors who accepted a conditional pardon as preferable to execution. Naturally, with these sorts of benefits to be gained, pardons were widespread.

Natural Law Arguments Against Pardon

It would be a mistake, however, to infer from the wide use of the pardoning power during this time that the right of a monarch to pardon had been fully established by philosophical argument. In reality, as the power to pardon grew, growing along with it was a religion-based argument against any pardoning power resting in a monarch.

The argument was rooted in the belief that in punishing lawbreakers, the state is carrying out God's will. The argument proceeded like this: Since all government comes from God, every crime is an offense against God. Moreover, God punishes those who offend against Him. It follows that whenever a magistrate pardons a criminal, he is usurping God's role and frustrating God's plan that those who sin should suffer.[20] The central premises of the argument were set forth by St. Paul to the Romans:

> Since all government comes from God, the civil authorities were appointed by God, and so anyone who resists authority is rebelling against

God's decision, and such an act is bound to be punished.... The authorities are there to serve God: they carry out God's revenge by punishing wrongdoers.[21]

"All government comes from God," the argument's first premise, meant different things to different people. For obvious reasons, it was an expression popular with kings, most notably England's James I, who said, "Kings are called Gods by the prophetical King *David*, because they sit upon GOD His Throne in the earth."[22] James I proceeded to use his divine connections to justify his exercise of power.

For St. Thomas Aquinas, writing in the thirteenth century, and for later natural law philosophers, St. Paul's premise was understood somewhat differently. For them, the premise was that human laws are derived from the divine law through the medium of divinely implanted human reason and natural law.[23] Just as God, when He created rocks and trees, created also the laws that they are bound to follow, God, when He created human beings, gave them laws to follow and the ability to learn what those laws are. The just state enacts and enforces these "natural" laws. The state is God's surrogate, the means God uses to enforce His will in the realm of human actions. Thus, a person who breaks the civil law sins against God.

"Such an act [sinning against God by breaking the civil law] is bound to be punished," St. Paul's argument continued. For natural law philosophers, the world provided strong evidence that God wants people who violate His law to be punished; that much is revealed by observing the physical world. God arranged the physical world so that whoever breaks divine laws governing physical objects is punished. People who defy the natural order of things by walking off a cliff, for example, are "punished" by the pain of a crushed spine. People who defy moral law by a night of drunken debauchery wake up in the morning with a headache sure to remind them of divine wrath. G. W. Leibniz saw it this way:

> [T]his city of God, . . . a moral world within the natural world. . . . God, the Architect, satisfies in all respects God the Law-Giver, that therefore sins will bring their own penalty with them through the order of nature, and because of the very structure of things, mechanical though it is. And in the same way the good actions will attain their rewards in mechanical way through their relation to bodies, although this cannot and ought not always to take place without delay. Finally, under this perfect government, there will be no good action unrewarded and no evil action unpunished; . . . [24]

Just as the order of nature enforces God's will concerning the physical limitations of human action, the civil state becomes God's tool for the enforcement of His will in the realm of political action. The general method of "divine administration" is to give mortals the capacity to foresee what enjoyments or sufferings will be the consequences of their actions. The administration of government works in the same way. By means of laws, governments annex pleasures to some acts and pain to others and forewarn citizens of these consequences of their possible acts. When the laws are nevertheless violated, the state imposes the promised punishment. Thus, St. Paul's conclusion: "The authorities . . . carry out God's revenge by punishing wrongdoers."

However, the assignment of pleasures and pains is not random. Under a perfect government, good actions are rewarded and evil actions are punished. When God "gets even," what is evened out is the amount of righteousness and reward, or pain and wickedness. As Bishop James Butler explained,

> Moral government consists, not barely in rewarding and punishing men for their actions, which the most tyrannical person may do; but in rewarding the righteous, and punishing the wicked; in rendering to men according to their actions, considered as good or evil. And the perfection of moral government consists in doing this . . . in an exact proportion to their personal merits or demerits.[25]

The government that consistently proportions punishment to wickedness is a just government. The God who does so perfectly is a just God.

That said, how do pardons fit it? If a society's laws are God-given, then the rulers of that society are carrying out God's commands in punishing. By connecting sin to suffering, they are playing their part in the divine ordering of the universe.[26] Punishing those guilty of sin is not something that the agents of God can decide to do or forbear. God's justice is at stake. If the state has a duty to do as God commands, then it has the duty to punish those who have sinned against Him.

While God has authorized civil governments to punish for Him, He cannot authorize them to forgive for Him. One can only forgive a wrong that has been done to oneself. Thus, a civil government can show no mercy to those who have broken the law and thereby offended against God. At the same time, however, God can forgive and show mercy to lawbreakers, and individual mortals are, in fact, urged to forgive those who trespass against *them*. But, when God wants to get even—through the procedures of mortal governments—mere mortals must not stand in the way.

Conclusion

Whatever its success among scholars, this argument did not sway those who held the pardoning power, as subsequent history testifies. On the Continent, the abuse of the pardoning power by the church grew to a dramatic climax. The sale of indulgences—documents granting divine forgiveness for sins—was one of the abuses that led Martin Luther to break with the Catholic Church. The resulting Reformation, the splintering of the Holy Roman Empire, and the political turmoil that followed can be laid at that door. In England, the struggle for who would control the privilege of abusing the pardoning power developed into a three-way tug of war (King, Church, Parliament) that spanned a half-dozen centuries. Not only wars and assassinations and executions—not only life-destroying acts—have determined the course of history. Misplaced mercy and extravagant pardons have played important parts as well.

This short chronicle of pardons before the Enlightenment is a portent of things to come. The pardoning power was necessary to soften the harshness and correct the injustice of the criminal law, but its abuses were frequent and cruel. The pardoning power necessarily operated outside of the law; however, its actual use was often lawless in the worst sense—arbitrary and corrupt. It was understood to be a God-like act of grace, but it usurped God's role in human affairs. The pardoning power, no less in the past than in the present, is paradoxical—a practical and philosophical puzzle.

2

Eighteenth-Century Reactions Against Pardon

Eighteenth-century Europe was the setting for reformist attacks against the great monarchies, attacks both military and philosophical. The monarchies were guilty of many abuses; among the most flagrant was their abuse of the pardoning power. They used the pardon frequently—to fill their treasuries, to populate their colonies, to win powerful friends. This abuse did not escape the attention of the political philosophers. The great treatises on government and the rights of citizens published by philosophers during this time usually had something to say about pardon, and usually it was critical.

Criticisms of Pardon

Some of the objections to the abuse of pardons were based primarily on practical considerations: monarchs could not expect to control crime if they continued to pardon so many criminals. In England, Henry Fielding joined a chorus of voices pointing out that pardons undermined the effectiveness of punishment. In his "Enquiry Into the Cause of the Increase of Robbers," Fielding wrote,

> Though mercy may appear more amiable in a magistrate, severity is a more wholesome virtue; nay, severity to an individual may, per-

haps, be in the end the greatest mercy. . . . The terror of the example is the only thing proposed, and one man is sacrificed to the preservation of thousands. If therefore, the terror of this example is removed (as it certainly is by frequent pardons) the design of the law is rendered totally ineffectual; the lives of the persons executed are thrown away, and sacrificed rather to vengeance than to the good of the public. . . . This I am confident may be asserted, that *pardons brought many more men to the gallows than they have saved from it.*[1]

For this reason, if for no other, Fielding said, an excess of pardons was a foolish self-indulgence.

But there were other reasons to oppose pardons. The Enlightenment philosophers were supporters of representative governments of one sort or another, and the place of pardons in a republic posed several problems.

First, who should hold the pardon power when powers are separated among several branches of government? Given the separation of powers, a President would have no more right to pardon than a federal jury would have the right to conduct a war. However, pardon power would allow a President to step across that separating line and nullify any judicial decision, thwarting legislative intent in the process.

There was another problem: In a monarchy, a crime is a crime against the King, who alone has the power to pardon. But in a democracy, a crime offends against the people. Who can pardon a crime, then, except the people themselves? In France, for example, Montesquieu made it clear that he had no objection to pardons in principle. In fact, they could be useful as a way of making the punishment fit the particular crime. But in a republic, Montesquieu declared, there could be no pardon. In a republic, the power to punish rested with the people, and a pardon would unjustifiably tamper with their decisions.[2]

Blackstone concurred: "In democracies . . . this power of pardon can never subsist, for there nothing higher is acknowledged than the magistrate who administers the law."[3]

Accordingly, the power to pardon was one of the institutions swept away by the French Revolution in 1789. For ten turbulent years, all pardons and commutations, all acts "tending to impede or suspend the exercise of criminal justice,"[4] were abolished. Why should the executive have the power to frustrate completely the enforcement of laws passed by the newly powerful legislature? And what use was the power to pardon, when sentences were to be pronounced by the members of a wonderful new institution, the jury—sensible, honest people who could be trusted to make the right decision in the first place? Eventually, France found that it could not get along without some machinery for clemency and made do with a variety of *ad hoc* procedures, until a pardoning power was formally granted to the First Consul in 1802. Since that time, as Leslie Sebba has pointed out, no other regime in the world has been without a clemency power of some sort. "The experiment of abolition was not only short-lived," Sebba wrote, "but also unique, never to have been repeated."[5]

The Pardoning Power in the New United States

The experience of the United States, the new republic creating itself on the other side of the Atlantic, was somewhat different from the European experience. Through the legacy of the great British jurists and political philosophers, and through a more bitter legacy left by their own experiences with royal governors possessed of the King's pardoning power, the colonists inherited an ambivalence about the place of pardon in a republic. Many colonists agreed with Montesquieu: If a pardon was to be understood, as it was in England, as an act of grace, a personal, man-to-man act of forgiveness, then there could be no executive pardoning power in a democracy, where a crime is an offense against the people, not an affront to the King.

Despite this argument the representatives assembled in Philadelphia in 1787 agreed on article II, section 2 of the U.S. Constitution after relatively little discussion. They wrote that the President

> shall have Power to grant Reprieves and Pardons for offenses against the United States, except in cases of impeachment.

Why would the Framers have agreed to a pardoning power that was King-like in concept and in scope? One possible explanation is that the Framers were more inclined to see the pardon as an instrument of law enforcement than as an act of grace.[6] Thus, they were more concerned with making the pardon power work than with making it conform to philosophical presuppositions about democracy. Since they wanted to empower the government sufficiently to execute the laws, they were more concerned about who could wield the pardon power most effectively than to whom the pardon power rightfully belonged.

This explanation is consistent with the arguments recorded by James Madison at the Constitutional Convention and the arguments used by Alexander Hamilton in defense of article II in *The Federalist*. Both reinforce the image of pragmatic men determined to guarantee an effective government even at the cost of compromising democratic principles.

The following is my paraphrase of the objections that were raised, and how they were answered:

Objection: *The pardoning power should be limited to pardons after conviction.*

Answer: Not so. A pardon before conviction might be needed to obtain the testimony of accomplices, in forgery cases, for example.

Objection: *The power should be granted to the legislature rather than to the President.*

Answer: This suggestion is ill-advised for several reasons. Suppose a President had sent a spy to ingratiate himself with a foreign enemy. The spy, returning, could find himself in difficulty. Only the President could pardon his secret agent, because only the President would know he was a spy.

Besides, legislatures are too often guided by passion. They cannot be trusted to use the pardoning power wisely.

Moreover, legislatures are not always in session. A person could be hanged before the legislature could be convened. And even when in session, legislatures work slowly. They might miss the chance for the best use of pardon. Hamilton wrote that

in seasons of insurrection or rebellion, there are often critical moments, when a well-timed offer of pardon to the insurgents or rebels may restore the tranquility of the commonwealth; and which, if suffered to pass unimproved, it may never be possible afterwards to recall.[7]

Objection: *Perhaps the pardoning power is too great for any one person, and should be shared in some way with the Congress.*

Answer: No, the best way to make sure that people act in a responsible way is to hold them to account as individuals. A legislature could avoid blame for the irresponsible use of pardon by dividing responsibility. Hamilton:

The reflection that the fate of a fellow-creature depended on his *sole fiat*, would naturally inspire scrupulousness and caution;

the dread of being accused of weakness or connivance, would beget
equal circumspection, though of a different kind.[8]

To my ear, these do not sound like the arguments of philosophers, debating fine points of justice and mercy. These are the views
of practical people, at pains to make a government that works, even
if that means granting full pardon power to the President of a democracy. In all of the record, there is only one mention of the relation
of pardon to considerations of justice or liberty. This comes from
Hamilton:

> The criminal code of every country partakes so much of necessary
> severity that without an easy access to exceptions in favor of unfor-
> tunate guilt, justice would wear a countenance too sanguinary and
> cruel.[9]

When all was said and done, the pardoning power remained as
it had been approved by the Constitutional Convention, in the hands
of the executive—not so that he could partake of the divine virtue
of forgiveness, but so that he could enforce the law.

Meanwhile, on the far side of the Appalachian Mountains, events
were unfolding that would put Hamilton's practical ideas to a practical
test: Could a well-timed offer of pardon restore tranquility? In home-
made stills, Pennsylvania farmers distilled their grain into whiskey,
a profitable product far easier to cart to market than corn-on-the-
cob. In 1794, several hundred of them violently objected to the federal
revenuers coming around and collecting a tax on their livelihood.
They burned down the house of one revenuer and tarred and feath-
ered others. With President George Washington at its head, an army
marched across the mountains. The Whiskey Rebellion ended as the
rebels ran off the field. Washington pardoned them all:

> For though I shall always think it a sacred duty to exercise with
> firmness and energy the constitutional powers with which I am vested,
> yet it appears to me no less consistent with the public good than it is
> with my own feelings to mingle in the operations of Government every
> degree of moderation and tenderness which the *national justice and
> safety may permit.*[10]

The President had put his finger right on two of the central problems
in the theory of pardon. The Constitution charged him with carrying
out the law of the land, and the same document gave him the right

to permit some who broke the law to do so with impunity. Merciful treatment of offenders is good in itself, and yet it is often unjust and unsafe.

Kant's Critique of Pardon

One year later, Immanuel Kant published, in *The Metaphysical Elements of Justice*,[11] what is probably the most notorious polemic against pardons. Kant argued that the members of the community, being the ones who are injured by a crime, are the ones who have the right to punish offenders. Rulers who grant pardons for crimes—other than those that injure them personally—usurp the rights of the subjects. Moreover, in an ideal state, the members of the community, through official authorities, have a categorical moral obligation to punish those who have committed crimes. To fail in the duty to punish—as they would if they pardoned an offender—is a breach of moral law.

Kant's opinion about pardons attracted attention, negative and positive and puzzled, and still does today. A measure of its unpopularity is the vigor with which some of Kant's interpreters have tried to explain it away. Some commentators on Kant have tried to pass off his view of pardon as the rantings of a usually serious man. H. von Hentig, for example, marveled in 1937 at how "even a mind so well trained in philosophy as Kant's, lost itself in atavistic emotional impulses, as soon as it had to do with matters of penal law."[12] James Heath primly wrote in 1963 that "the zeal with which Kant advocated retributivism may excuse a doubt whether he was not himself much influenced on this matter by rather primitive emotions."[13] Perhaps the most thoughtful contemporary response came from Jeffrie Murphy, who caught Kant in serious self-contradictions and admitted that he "despair[s] of finding a consistent Kantian view of crime and punishment."[14]

It may be possible to interpret Kant's opposition to pardon historically, as a reaction against the misuse of pardon rampant during his lifetime.[15] The daily news may have reinforced Kant's view that the right to pardon belongs to the people. It is preferable, however, to try to understand Kant's view of pardon as originating in his theory of justice rather than in the sourness of his spirit or the corruption of his King. To do so, it is necessary to explain why Kant thought that criminal punishment is justifiable.

This is a job made very complicated, as Murphy pointed out,[16] by how difficult it is to find a single set of Kantian views about punishment and to understand what Kant meant even when he was being consistent. What follows is not so much an explanation of Kant's general theory of punishment (which may or may not exist), but one possible construction based on passages from *The Metaphysical Elements of Justice*, a construction that ignores the more mellow, consequentialist Kant of some of his other work.

Kant on Punishment

Writing in *The Metaphysical Elements of Justice*, Kant argued that punishment of those who have committed a crime is a categorical obligation binding upon the community through its officials. In a famous passage, Kant wrote with passion and clarity:

> The law concerning punishment is a categorical imperative, and woe to him who rummages around in the winding paths of a theory of happiness looking for some advantage to be gained by releasing the criminal from punishment or by reducing the amount of it. . . .[17]

There are no exceptions; even if a community decided to disband and its members scattered throughout other nations, it would first have to make sure that all its criminals had been punished, that all its murderers had been executed. By doing so, the members of the society would make sure that all got what they deserved, thus making sure that the members of society were not themselves guilty of injustice.[18]

Several arguments led Kant to these conclusions. Of primary importance was Kant's conviction that the failure to punish constitutes an injustice. For Kant, justice was the principle by which each person is obliged to act externally in such a manner that the free exercise of his will may be able to coexist with the freedom of all others, according to a universal law.[19] A just state is a community of citizens who agree to make sacrifices by obeying laws that limit their external freedom, with the understanding that others will make the same sacrifice. All thereby equally gain the protection of their liberties.[20] This principle of equal freedom is important. Kant illustrated it by the pointer on the scale of justice.

Equal protection of freedom under the law is the essence of the social contract. All members of the society are co-legislators, creating a state with the power to protect their freedom from external constraint. They create the law under which they will live, including the penal law. Then they submit themselves, along with all other members of society, to the criminal laws and penalties. When all obey the law, all enjoy freedom equally, and the pointer on the scale of justice records a just equilibrium.

A person who disobeys the law does the other members of the community a comparative injustice by taking liberties that he is unwilling and unable to grant to others. Moreover, his external acts interfere with others' free exercise of their wills. This is acting unjustly. The pointer on the scale of justice is awry, as the added weight of the liberty a criminal takes for himself out-balances the reduced freedom enjoyed by those who continue to obey the laws.

Punishment limits the freedom of a person who has violated the law and thereby removes the injustice of his act. As two wrongs can sometimes make a right, taking away the freedom of someone who has taken away the freedom of others is a kind of justice. And serving justice is the proper role of the state. Accordingly, the state has not only the right to punish, but an obligation to punish, when punishment is required to restore justice. Justice is the "general justifying aim" of punishment.

Furthermore, in *The Metaphysical Elements of Justice* (although not elsewhere),[21] Kant stood squarely against the utilitarian view that was soon to swamp retributivism in a storm of reform. It is never right to punish someone for the benefit of someone else, or for the good of society, Kant said. Such a plan manipulates human beings as if they were objects to be used as tools for another's happiness. People may not be treated as objects only; their standings as human beings protect them from that.[22]

Just as the principle of equal freedom provides the justification for punishment, Kant insisted, it provides the measure of punishment. This is the law of retribution, compensating for the offense by returning like for like. The person who steals must lose his property. The person who kills must lose his life. Adjustments must be made, of course, to make this a practical effort. But nothing less than like for like will restore equality, nothing less will be weighty enough to move the pointer of justice back into balance.

Given this construction of a Kantian theory of punishment, how does the executive pardon power fit in?

Kant's Objections to Pardon

Once punishment is seen as a kind of justice, it is a small step to infer that pardons, which block or interfere with punishment, will be seen as a hindrance to justice. And indeed, with some important exceptions, Kant took a dim view of pardons. Although his arguments are not fully developed or consistent, it is possible at least to find several ways pardons are unjust according to a Kantian view of punishment.

According to one interpretation, Kant seemed to suggest that a pardon violates the rights of the law-abiding citizen. This interpretation presupposes that punishment can be understood as a sort of promise-keeping and a legal system as a system of promises.[23] A statute making an act of a certain kind illegal and affixing a penalty to its violation can be viewed contractually: *i.e.*, "If you do *[what the law forbids]*, then the state will *[punish you as the law prescribes]*." The advantage of a contract in a business transaction is that the parties to a contract can predict with accuracy the consequences of their actions. In addition, they have a means of ensuring that those expected consequences are forthcoming. The point is the same when laws are viewed as contracts. The law makes a promise to the law-abiding citizen (if you obey the law, you will not be punished) and to the criminally inclined (if you disobey the law, you will be punished). The very promise-making institutes an obligation to punish when the violations occur.

If someone disobeys the law and the state fails to punish him, then the law-abiding citizens have a legitimate claim against the state. Their right to have criminals punished has been infringed. It follows that a sovereign cannot decide not to punish.

The sovereign can, however, pardon an injury to himself. This does not turn out to be much of a comfort to the criminal. According to Kant's political theory, virtually all crimes—public and private— cause injuries to members of the community rather than to the sovereign. All public crimes hurt all members of the community. Private crimes hurt the victims. Only in a very rare case would a crime injure only the sovereign, as, for example, if the sovereign was defrauded of his personal fortune. Hence, with only one class of infrequent exceptions, the sovereign may not pardon offenders.

A passage in *The Metaphysical Elements of Justice* indicates that this is indeed what Kant had in mind when he criticized the use of pardon:

The right to pardon a criminal, either by mitigating or by entirely remitting the punishment, is certainly one of the most slippery of all the rights of the sovereign. By exercising it he can demonstrate the splendor of his majesty and yet thereby wreak injustice to a high degree. With respect to a crime of one subject against another, he absolutely cannot exercise this right, for in such cases exemption from punishment constitutes the greatest injustice toward his subjects.[24]

There is another possible explanation for why pardons are generally unjust on Kantian grounds. Pardoning a criminal is unfair to the law-abiding members of society because it allows the criminal to enjoy an advantage unfairly won and undeserved. H. J. McCloskey[25] has offered this argument, which reflects a Kantian point of view[26]: "Mercy without good reason is a form of immorality."[27] If the decision is made not to punish a criminal, then there must be a system of rewards for those who obey the law. Otherwise, people would end up getting more or less than they deserved, and the state would be violating its duty to protect equal liberties. "There is an evilness about unpunished crime,"[28] because it is a kind of injustice—an injustice against which law-abiding citizens have a right to be protected.

So if there is going to be punishment at all, it must be administered unfailingly and impartially. It follows that pardoning someone who has broken the law is unjust, given a generally Kantian view of the moral and political justification of punishment. Or so it would seem.

Kantian Grounds for Pardoning

Despite the harshness of his theory of punishment, Kant was not unrelievedly uncompromising. He himself suggested several circumstances that could justify pardon.

A first possible ground for pardoning rests on the distinction between an ideal and an imperfect society. While punishment is a categorical imperative in an ideal society, Kant said, it is only a hypothetical imperative in an imperfect society—certainly an apt description of both Kant's society and our own.[29] Murphy explained why this might be so: When the members of a society are of approximate equality and when the rules that they must follow have been commonly accepted and protect everyone equally—as would be the case in an ideal society—then it makes sense that a breach of the law is unjust and deserves punishment. Pardoning under those circumstances would perpetuate the unjust state of affairs created by the crime.

But suppose that the citizens do not enjoy equal liberty, that the laws do not protect all equally. In that case, a disadvantaged person committing a certain crime would not be taking any liberties that do not belong to him by right. Hence, the justification for punishment is defeated, and a pardon in such a case can not be criticized as unjust.

Writing to J. B. Erhard, Kant said,

> In a world of moral principles governed by God, punishments would be categorically necessary (insofar as transgressions occur). But in a world governed by men, the necessity of punishments is only *hypothetical*, and that direct union of the concept of transgression with the idea of deserving punishment serves the ruler only as a prescription for what to do.[30]

Kant himself suggested another circumstance under which pardons might be justified—when the crime is encouraged by the state whose laws forbid it. Kant provided two examples, infanticide and dueling, criminal acts that society encouraged by fostering a foolish regard for honor.[31]

Finally, a discussion of Kant on pardons would not be complete without noting a passage in *The Metaphysical Elements of Justice* in which Kant argued that commuting a sentence of death to deportation is at least excusable on consequentialist grounds. Kant explained that, if a murder was committed by an enormous number of people, then,

> the state would soon approach the condition of having no more subjects if it were to rid itself of these criminals, and this would lead to its dissolution and a return to the state of nature.... Since a sovereign will want to avoid such consequences and, above all, will want to avoid adversely affecting the feelings of the people by the spectacle of such butchery, he must have it within his power in case of necessity... to assume the role of judge and to pronounce a judgment that, instead of imposing the death penalty on the criminals, assigns some other punishment that will make the preservation of the mass of the people possible, such as, for example, deportation.[32]

This is an argument that is hard to reconcile with Kant's repudiation of consequentialist justifications. Indeed, Kant's argument would seem to be a good example of "rummag[ing] around in the winding paths . . . looking for some advantage to be gained . . . by reducing [punishment],"[33] which he so strongly derided.

This apparent inconsistency could be understood if Kant considered death and deportation to be punishments roughly equal in severity: Then there would be no injustice in changing a death sentence to a sentence of deportation as a matter of convenience, just as there would be no injustice in selecting hanging over beheading.[34] But in that case, the commutation would no longer be a kind of pardon.

With these exceptions and despite some inconsistencies, Kant made it clear enough that the sovereign who pardons crimes against the people is a usurper, doing his people an injustice by passing up the opportunity to correct the injustice brought about by a crime.

3

The Utilitarians' Position

In his opposition to pardon, Kant had an ally from an unexpected quarter. He was Jeremy Bentham, the great British reformer and utilitarian philosopher of the late eighteenth century. Writing from a moral point of view diametrically opposed to Kant's, Bentham came down hard on pardon as it was then practiced, proving (if there was ever any doubt) that politics does indeed make strange bedfellows.

Bentham was appalled by the hideous penal practices of his day: the disease-ridden prison ships floating dockside in London, the gibbets with their dangling thieves, and the still crime-filled city. He set out to find a better way. His answer was not fewer punishments, but fewer pardons and a penal system that used the prospect of humane but inexorable punishment to deter crime. On the subject of pardons, Bentham spoke his mind with characteristic forcefulness:

> From pardon power unrestricted, comes impunity to delinquency in all shapes: from impunity to delinquency in all shapes, impunity to maleficence in all shapes: from impunity to maleficence in all shapes, dissolution of government: from dissolution of government, dissolution of political society.

The Principle of Utility

Bentham's masterwork, *The Principles of Morals and Legislation*, was based on one overriding moral claim: that one ought always to act in a way that increases happiness and reduces pain. This principle provides a standard for judging individual acts and for judging legislation and government institutions. Any law or practice should promote the interests of the community. That is to say, it should add to the sum total of pleasures in the community and diminish the sum total of pains.

This is the famous "principle of utility." Bentham defined it this way:

> By the principle of utility is meant that principle which approves or disapproves of every action whatsoever, according to the tendency which it appears to have to augment or diminish the happiness of the party whose interest is in question.... A measure of government... may be said to be conformable to... the principle of utility, when in like manner the tendency which it has to augment the happiness of the community is greater than any which it has to diminish it.[1]

The Utilitarian Justification for Punishment

The business of government is to promote the happiness of society. It does so by means of laws that prohibit actions that cause "mischief," Bentham's wonderful word for "harm." A person who breaks a law does *primary* mischief to the person who is the victim of the crime and to those who care about or depend upon the victim. In addition, the criminal does *secondary* mischief through the fear aroused in people that they will suffer the same harm—and through the very real danger that they *will* suffer that harm as others follow the criminal's example.[2]

The business of punishment is to control actions so as to prevent such mischief. Punishment does this by reforming or disabling criminals, those who might otherwise commit additional crimes. But by far its most important effect is its influence on those who are deterred from criminal acts by the suffering imposed on those who broke the law. By reducing crime in these ways, punishment prevents more unhappiness than it causes.

Punishment "augments the happiness of the community" in one more way, Bentham noted. Victims and others get pleasure from the pain suffered by those who have injured them. While vindictive satisfaction is never strong enough to outweigh the criminal's pain—and thus does not itself provide a justification for punishment—punishment justified on other grounds ought to be imposed in a way that accommodates the victims' desire for revenge. Every little bit of pleasure counts in the hedonistic calculus.

The utilitarian justification for punishment can be stated more formally:

1. The state has the duty to achieve a specified object.
2. Laws are the instruments by which the state is to reach its object.
3. Infractions of the law frustrate the achievement of the object.
4. The state has the right to punish infractions of the law so far as this is necessary to achieve its object and within the limits established by the nature of its object.

Proposition 4 limits the use of punishment; if a punishment does more harm than good it cannot be imposed. "Every punishment which does not arise by absolute necessity . . . is tyrannical," wrote the eighteenth-century Italian criminologist and reformer Cesare Beccaria.[3] When does a punishment "not arise by absolute necessity"? When crime—and its resulting unhappiness—can be prevented by means less painful than punishment. And even if justifiable, a punishment may not impose more pain than it prevents. This is a bitter sort of "checking account": A given punishment puts money in the bank, in the currency of "pain prevented." The same punishment draws from the account, in the currency of "pain caused." The account must never be overdrawn.

It is important to note here what does *not* justify punishment. People can never be punished just because they have committed a crime. People can never be punished just because they deserve to be punished. All that, to a Utilitarian, is water over the dam, irrelevant information. Revenge *may* justify punishment, not because it evens the score, but because people often feel good when the score is evened. In sum, no justifications for punishment count unless they are facts about the future—the effect of punishing or not punishing on the public interest.

An eighteenth-century judge echoed this philosophy: "You are to be hanged not because you have stolen a sheep but in order that others may not steal sheep."[4] So also did the Marquis deJaucourt, writing in Diderot's *Encyclopédie*:

> Although, under the Law of Nature, punishment follows upon crime, it is clear that the sovereign ought never to inflict it unless someone will be benefited. To impose suffering upon anyone, simply because he has made another suffer, is an act of pure cruelty, con-

demned by Reason and Humanity.... In punishment ... there must always be in view either the welfare of the culprit himself, or the interest of the victim, or the advantage of the community.[5]

The Duty to Punish

With this emphasis on future interests and advantages—even on "the welfare of the culprit himself"—one might expect that the Utilitarians would benevolently welcome the chance to ease criminals' pain by easing their punishment. Yet this is not so. The same premises that establish the state's right to punish establish the state's duty to punish.

Laws and legal institutions must serve the public welfare. That is a primary obligation in the utilitarian view. The obligation to punish is secondary and instrumental, derived from the primary obligation to augment happiness. The obligation to punish can thus be understood as a hypothetical imperative: To the extent that a punishment furthers the object of the community, it is demanded by the precepts of morality. As DeJaucourt wrote, "The sovereign, by his office, is not only entitled, but indeed obliged, to punish crimes."[6]

But one might ask the Utilitarian, What is the harm in not punishing? Surely there are cases—old, sick men who have committed petty crimes, or ragged mothers with hungry babies—where the pain of punishment would be so great that punishment should be foregone. What possible harm could come from letting these people go home?

The utilitarian answer is surprisingly uncompromising. A penal system that hopes to deter crime cannot tolerate exceptions. Punishment deters crime not only in criminals themselves, by reforming or disabling them, but in others as well, by setting an example. The first sort of deterrence, specific deterrence, is important. But, because it affects the actions of so many more people, the second sort of deterrence is proportionally more important. A pardon in one case leads others to infer that they too may be able to commit a crime and escape punishment. Rather than being discouraged from crime by fear, they are encouraged to crime by hope that they will escape punishment. Thus, the deterrent effect of punishment breaks down, and subsequent punishments are wasted, as they inflict pain for no compensating return—a cruelty that cannot be justified.

In 1925, Oliver Wendell Holmes wrote:

> If I were having a philosophical talk with a man I was going to
> have hanged (or electrocuted) I should say, I don't doubt that your
> act was inevitable for you but to make it more avoidable by others we
> propose to sacrifice you to the common good. You may regard yourself
> as a soldier dying for your country if you like. But the law must keep
> its promises.[7]

The Utilitarians' efforts to guarantee the deterrent effect of punishment require them to judge not the merits of individual cases but
rather the merits of general rules. Individual *acts* of punishment are
not tested against the utilitarian requirement that each produce more
good than withholding punishment. The beneficial consequences of
punishment depend on its certainty. So it must be general *rules* about
punishing that are subjected to the utilitarian test. Individual acts of
punishment are then justified if they comply with utilitarian-justified
rules about punishing and pardoning. Contemporary students of
moral theory will recognize this as a shift between standards that have
come to be known as act-utilitarian and rule-utilitarian.[8]

A legal system should be humane and merciful, of course. It
should not impose unnecessary suffering, even on criminals. But, the
Utilitarians end up saying, a legal system should not make exceptions
to rules. Thus, the rules themselves must be humane, but the enforcement of those rules must be absolute. Beccaria suggested that
criteria for pardoning should be worked into the laws by legislating
with temperance, rather than left within the power of the magistrate:

> Clemency is a virtue which belongs to the legislator, and not to the
> executor of the laws, a virtue which ought to shine in the code, and
> not in private judgment. To show mankind that crimes are sometimes
> pardoned, and that punishment is not the necessary consequence, is
> to nourish the flattering hope of impunity, and is the cause of their
> considering every punishment inflicted as an act of injustice and oppres-
> sion. Let, then, the executors of the law be inexorable, but let the
> legislator be tender, indulgent, and humane.[9]

This adds up to a utilitarian presumption in favor of punishment.
The state ought to punish criminal offenders, unless it is generally
true that the evil of punishing is greater than the evil of pardoning
in cases of a given kind.

This conclusion was summarized by S. I. Benn, a twentieth-
century Utilitarian. He wrote that the establishment of legal guilt
overcomes the initial utilitarian presumption against causing delib-

erate suffering. Further, determination of legal guilt is the sole condition for the creation of a *prima facie* obligation to punish. The *prima facie* obligation may be defeated by other utilitarian considerations. But those other considerations must take the form of general grounds for pardoning. And any conditions for pardoning must meet the following formal criterion: "To recognize it as a general ground for waiving the penalty would not involve an otherwise avoidable mischief to society *greater* than the mischief of punishing the offender."[10]

General Grounds for Pardoning

If the state has a *prima facie* obligation to punish all whose legal guilt has been established, it becomes important to learn what facts in general have the moral credentials to defeat the obligation to punish. Such facts turn out to be the conditions for justified pardons.

Bentham's Pardon Conditions

Bentham carefully tested a list of crimes against his happiness principle and presented a catalogue of cases in which punishment ought not to be imposed.[11] Punishment ought not to be inflicted (1) where it must be inefficacious, (2) where it is groundless, (3) where it is needless, and (4) where it is too expensive.[12] Bentham briefly explained what he meant by each kind of case.

1. When punishment would be ineffective in deterring crime, it is not permissible. Punishment or the threat of punishment would not be a deterrent in cases of *ex post facto* laws or secret laws. Likewise, there is no point in punishing infant, insane, intoxicated, or incapacitated persons. Insofar as lawbreakers such as these do not have control over their actions, their crimes could not have been deterred. Nor will punishment reform any of these. Even more important, it is unlikely that other, similar people will draw dangerous conclusions from this sort of immunity and be encouraged to commit crimes, or so Bentham believed.

2. Further, punishment should not be inflicted when it is "groundless," Bentham wrote. That is, punishment ought not to be inflicted when the crime in question has not caused any "mischief" for a variety of reasons. It might be that the victims consented to

having the crime perpetrated against them. Or the case might be one where a crime is committed to prevent a greater harm; pardon is called for when, for example, someone 'steals' the sheets from a clothesline to bind the wounds of a traffic victim. In addition, certain harms that can be completely repaired (embezzlement, for example) are pardonable.

3. Sometimes a punishment is needless. Education and social reorganization are often effective in stopping crime. In such cases, they should be preferred to punishment, since they are 'cheaper' in the coin of pleasure and pain.

4. There are certain types of cases in which a punishment would cause harm greater than the harm of not punishing. First, if a large segment of the population commits a crime, it might be prudent to forego punishment. The amnesties provided Vietnam War draft evaders and illegal aliens are contemporary cases in point. Second, it could be that some people who stand to be punished are in a position to render such an important service to the community that their imprisonment would be a great loss and would therefore be unjustified. Third, when punishment of a given criminal would endanger a nation by raising the ire of a foreign power, it should be foregone.

Although Bentham did not mention this possibility, deJaucourt pointed out that it may be unjustifiably 'expensive' to punish some particularly tantalizing crime if the atrocity occurred in secret.

> If a crime has been secretly committed, so that it is known to very few, there is not invariably a need to punish it—sometimes, indeed, to do so would be very dangerous—as the result of punishing it must be to publish it also, and some there will be who abstain from a wrong action rather through ignorance of the ways of wickedness than through awareness and love of virtue.[13]

For all of these reasons, Utilitarians agree that the duty to punish must be overridden in a limited collection of cases. These cases have characteristics that defeat the justification for punishment; in cases of these kinds, punishment does more harm than good.

For Utilitarians, the punishment does not have to fit the crime, nor does it have to fit the criminal. The punishment must fit only the needs of society. Legislation should be carefully drawn to tighten the fit, to make sure that no punishment is imposed that does not accomplish some future societal good.

Defining the place of pardons in this tight a system of punishment is a very difficult thing to do. Perhaps it is no surprise that the Utilitarians have not entirely succeeded.

Inconsistencies in Bentham's Pardon Conditions

When Bentham drew up his list of types of cases where punishment should be not inflicted, he may have meant to say that under the four conditions, no crime was committed. If so, he was in effect following Beccaria's recommendation that laws be drawn so carefully and so humanely that a pardon is never needed. On the other hand, he may have meant that these are cases in which a crime has been committed but should not be punished—an interpretation closer to his own wording. If so, it is important that the pardon conditions be consistent with utilitarian principles. It is my view that two of them are not.

According to Bentham, punishment should not be imposed if it is *inefficacious*; that is, if it is not likely to prevent crime. His examples are interesting: many of them are cases in which, in modern parlance, lawbreakers acted without criminal intent. Bentham's list includes people who did not intend to do harm because they were insane or intoxicated, people who made a mistake of fact because of severely limited intelligence, and people who lacked physical control over their actions. These sorts of people do clearly have one thing in common— they cannot be blamed. But Bentham, a faithful Utilitarian, cannot care about *that*. All that counts for him is summarized in his claim that punishing people like this would do no good. But I think there is reason to believe that Bentham is mistaken about that.

Insane people who commit crimes are dangerous, in particular *because* they are insane. People of subnormal intelligence who hurt others are dangerous, all the more so because they do not understand the risks they take. Preventing them from doing more damage by incarcerating them may be cruel, but there is no denying it would be efficacious. Moreover, their crimes set a particularly strong example both among irrational people who are easily influenced by example and rational people who might be encouraged to feign insanity to escape punishment. So, putting people such as these in jail or in places of safety would be efficacious indeed—preventing both primary and secondary mischief, to use Bentham's language. Whether this is called 'incapacitation' or 'punishment' does not make much difference; Bentham's theory of punishment blurs the distinction.

Why did Bentham exempt them from punishment? Perhaps he underestimated the dangers posed by nonresponsible people, an empirical miscalculation. Or, more likely, he was reluctant to punish people who cannot be blamed. While commendable, this is a moral squeamishness that does not follow from the utilitarian view.

A twentieth-century British Utilitarian, Lady Barbara Wootton,[14] was far truer to her utilitarian colors. Recognizing that "in the modern world . . . more damage is done by [those who do not intend it] than by deliberate wickedness,"[15] Wootton argued that criminals who cannot be held responsible should be 'punished' (or incapacitated) nonetheless, insofar as punishment can restrain them from doing more harm. It seems to me that if Bentham had been consistent, he would have been forced to take Lady Wootton's position on nonresponsible offenders.

Moreover, I do not understand how the *groundless* pardon condition follows from utilitarian principles. Bentham said that a good reason for not punishing is that a crime—because of its particular circumstances—has not done any harm. How can this be a pardon condition when his theory of punishment has specifically made the *nature* of the crime irrelevant to a judgment of the punishment to be imposed? All that is to be considered is a prediction of the consequences of punishing or failing to punish. Thus, the harm that the crime caused in the past cannot count in the decision-making process. A Utilitarian fits the punishments to the requirements of the community. To fit the punishment to the crime, as this condition would do, requires different principles altogether.

However, that Bentham bends his principles in such a way is interesting. For all *but* Utilitarians, the harm done by the crime is important to a decision about punishing. This is because the amount of harm done by a crime serves as a measure of how much punishment the criminal deserves. The utilitarian calculus, however, is an attempt to get away from such abstract notions as desert, to make punishment decisions solely on the basis of what will best serve the community. That Bentham allowed for the elimination of punishment when a crime has not done harm is a testimony to the strength of the conviction that a person who has done no harm does not deserve to be punished, regardless of the effects of the failure to punish on the populace. But one cannot have this conviction and the greatest-happiness principle as well.

Benevolence and Pardon

The biggest disappointment about the utilitarian handling of pardon is not its inconsistency but the fact that it can seldom justify making exceptions to ease individual suffering.

The political theory based on the greatest-happiness principle forbids an executive from granting a pardon simply to reduce the misery of a single criminal. A desperately ill prisoner, his starving children, his grieving wife—their happiness counts for little, swallowed up by the interests of the community as a whole.

The right of punishing, as Beccaria pointed out, belongs to society as a whole, or to the sovereign, or to those chosen to protect the rights of citizens.[16] Although an individual may be the victim of a crime, it is against society in general that a crime is perpetrated. So while a wronged individual may forgive the person who wronged him (an act of good nature and humanity), he cannot remove the necessity of punishing the offender. A victim can lay down his own right to punish, but he cannot forfeit the rights of society as a whole.

On utilitarian principles, what one ought to do—what justice requires—in regard to both the duty to punish and the duty to pardon, is defined by a general principle of benevolence to society in general, the duty to promote the greatest good on the whole. Thus, there is a proper punishment for every case, determined by weighing the consequences of the punishment. The considerations that might lead to a recommendation for pardon in other systems—the suffering of criminals or their loved ones—will have already been weighed by a utilitarian judge in determining the best sentence.

Thus, on utilitarian grounds, judges are not in a moral position to act mercifully; they must impose the sentence that will be most productive of good. Once that sentence is determined (by weighing all the relevant factors), the judges may not act mercifully and impose less than the best punishment. Alwynne Smart made this point:

> There is no significant sense in which the utilitarian can say, "I *ought* to do such and such, but special considerations persuade me to act differently on this occasion." For him, the statement, "I shall act mercifully" can *only* mean "I shall impose a penalty less than the one which will produce most good," which in turn can *only* mean, "I shall impose

a penalty less than the one which will produce most good because this action is the one which will produce the most good."[17]

There can be no exceptions for the sake of individual happiness in a system of punishment devoted to achieving the greatest happiness for the greatest number.

4

The Nineteenth Century: Pardon and the Right to Punishment

There was one last furious philosophical blast at pardons in the nineteenth century, before the topic faded away for a hundred years as philosophers turned their attentions to other things. The argument against pardon was based on G. W. F. Hegel's premise that criminals have a right to be punished. Meanwhile, the U.S. government was setting about the grim business of establishing institutions of punishment and refining a theory of pardon through the slow process of practice and precedent.

Pardon and the Criminal's Right to be Punished

The last great opponent of pardon was Georg Wilhelm Friedrich Hegel, an early nineteenth-century German Idealist. His criticism of pardon was based on the view that a pardon violates a right possessed by every criminal—the right to be punished. The idea that criminals have a right to be punished no doubt looks strange to some people, particularly people in jail. It is a right that many people would be happy to forfeit.[1] But they should think twice. When a person gives up the right to be punished, much more is lost than a prison sentence.

Punishing people implicitly acknowledges them to be persons, moral agents making free choices.

The Right to be Punished; The Duty to Punish

A civilized society does not punish imbeciles, it does not punish wild animals, it does not punish trees—even when they do serious harm. It does not punish them because it recognizes that imbeciles, animals, and trees do not have the mental capacities to make their own decisions and to act in accordance with them. When a society punishes people, it implicitly acknowledges that they are not objects. They are rational beings acting freely. Pardoning them denies that and thereby violates their rights to be treated as persons.

Because Hegel's prose tends to be murky, I think the best statement of Hegel's view comes not from Hegel himself, but from an English political scientist, Ernest Barker:

> The assumption . . . made, in the act of punishment, that each agent is individually responsible, is a tribute to the principles of liberty and equality, and thereby to the ultimate and intrinsic value of human personality from which those principles are derived. Punishment is the black shadow of those principles and that value; but we respect them most if we recognize that they do, and must, cast a shadow. . . . If we respect responsibility, *we must respect the right of offenders to be punished for their offenses.*[2]

An exasperated prisoner might happily trade all this respect for a little freedom, but Hegel took the right to punishment very seriously indeed.

Hegel's argument can be paraphrased this way: A human being is a rational being. Thus, a criminal act is the direct result of a choice—not of an animal urge, not of an unhappy childhood, not of a blind impulse. A crime is the result of a conscious decision that acts of that sort are right. People have a right to institutions that respect them as rational beings and thus take their choices seriously. If a state takes a criminal choice seriously, it recognizes that the choice represents a universal principle that stands as a direct challenge to the state's rules and must be punished. In addition, punishing criminals takes their choices seriously by applying to their cases the universal principle embodied in their decisions to commit crimes.

Hegel wrote that

what is involved in the action of the criminal is not only the concept of crime, the rational aspect present in crime as such whether the individual wills it or not, the aspect which the state has to vindicate, but also the abstract rationality of the individual's *volition*. Since that is so, punishment is regarded as containing the criminal's right and hence by being punished he is honoured as a rational being. He does not receive this due of honour unless the concept and measure of his punishment are derived from his own act. Still less does he receive it if he is treated either as a harmful animal who has to be made harmless, or with a view to deterring and reforming him.[3]

For these reasons, Hegel claimed that "the administration of justice must be regarded as the fulfillment of a duty by the public authority."[4] That duty is the correlate of the criminal's right to be punished.

What is important about a criminal act is not that it breaks a rule forbidding it, Hegel argued, but that it denies the rightness of the rule it breaks. A murder, for example, causes two sorts of evil— the death of the victim and a more wide-ranging evil, the implication that murder is not wrong. A crime speaks a kind of universalizing language of action.[5]

This has two consequences. First, it is important to punish an offender in order to restore the principle which was infringed. "[T]o injure [or penalize] this particular will [of the criminal] as a will determinately existent is to annul the crime, which otherwise would have been held valid, and to restore the right."[6] If the crime is a denial of the rightness of the rule broken, then the negation of that denial restores the rule.

In other words, punishment does in the criminal law what reparations do in the civil law. Hegel argued by analogy that

> [in] so far as the infringement of the right is only an injury to a possession or to something which exists externally, it is a *malum* or damage to some kind of property or asset. The annulling of the infringement . . . is the satisfaction given in a civil suit, *i.e.*, compensation for the wrong done, so far as any such compensation can be found.[7]

A smashed carriage, for example, can be repaired or replaced. In a criminal case, however, the damage of consequence is to the principle contravened. The only injury that *exists* (as the smashed carriage exists) is the will of the criminal—the claim made by his crime that the principle forbidding his act is invalid, that his act is morally permissible. The punishment is said to restore the principle by doing violence to that will. Punishment annuls the particular crime.

But this is only part of the story. It is important to punish an offender because that is what the offender wills. By a criminal act, Hegel wrote, a criminal explicitly establishes a certain universal principle: that the criminal act, and others relevantly similar, are right. A highwayman, for example, by an assault and theft explicitly establishes the principle that it is right to rob others. But since the highwayman is a rational (that is, consistent) being, he realizes that the same principle should be observed relative to himself—that he should be similarly injured. That is what the punishment does: it mugs the mugger, murders the murderer, robs the robber. In Hegel's language,

> [Punishment] is the reconciliation of the criminal with himself, *i.e.* with the law known by him as his own and valid for him and his protection; when the law is executed upon him, he finds in the process the satisfaction of justice, and nothing save his own act.[8]

Respect for his principle, and thus his punishment, belongs to the offender by right. He has a legitimate claim to punishment, and the state has a corresponding duty to inflict it.[9,10]

The Hegelian Argument Against Pardon

How does this view of punishment argue against pardon? Pardons defraud offenders of their rights—the right to the recognition of their hostile wills and the right to be treated not as objects but as human beings.

Only punishing preserves the self-determination of offenders, because punishment is what they choose when they choose to commit crimes. Since persons have the right to be punished for their crimes, the state has not only the right, but the duty, to punish. Pardon, the failure to punish people who deserve punishment, is a moral failing, a breach of obligation, and a violation of right. The obligation to punish, Hegel believed, is absolute.

Pardon in Nineteenth-Century America

Ironically, it was a robber's insistence on his right to refuse a pardon that led to the landmark case of *U.S. v. Wilson*, which entrenched mercy or "grace" as the doctrinal keystone for almost a century of U.S. court decisions about pardon.

The U.S. Supreme Court: Pardon as an Act of Grace

The defendant, George Wilson, evidently had a lucrative career robbing the U.S. mail and endangering the lives of mail carriers. In 1830, this was a capital offense. The language of the indictment provides the flavor of the times:

> [Wilson] feloniously did make an assault, and him the said Samuel M'Crea in bodily fear and danger then and there feloniously did put, and the said mail of the United States from him the said Samuel M'Crea then and there, feloniously, violently and against his will, did steal, take and carry way, contrary to the form of the act of Congress in such case made and provided, and against the peace and dignity of the United States of America.[11]

Wilson entered a plea of guilty to a whole web of crimes and was sentenced to death for one of them. President Andrew Jackson, responding to high-level recommendations, pardoned him. Wilson was then brought to trial on a judicially separate but factually related charge.

To this second, related, charge, Wilson entered a plea of "not guilty." Knowing that the pardon implied the guilt that he now denied, Wilson waived his right to the protection of the pardon. At issue: Can Wilson refuse to accept a pardon? The court's answer: Yes, of course, just as one can refuse a gift.

Disregarding the strident philosophical objections to the traditional English concept of pardon, disregarding evidence that the Framers were not enthusiastic about English pardoning practices, Chief Justice John Marshall looked to English common law for a basis of decision.

> As this power had been exercised from time immemorial by the executive of that nation whose language is our language, and to whose judicial institutions ours bear a close resemblance; we adopt their principles respecting the operation and effect of a pardon, and look into their books for the rules prescribing the manner in which it is to be used by the person who would avail himself of it.[12]

Accordingly, Marshall pronounced pardon to be an act of grace:

> A pardon is an act of grace, proceeding from the power entrusted with the execution of the laws, which exempts the individual, on whom

it is bestowed, from the punishment the law inflicts for a crime he has committed. It is the private, though official act of the executive magistrate.... [13]

Since it was a kind of gift, the pardon could be refused. The decision saved Wilson—but it almost killed off judicial interest in pardons. For the next hundred years, Presidents used pardons as they chose, having been given a pardoning power patterned after that of the English Kings, which was patterned after God's.

Nineteenth-Century Use of Presidential Pardons

The most famous pardons of the nineteenth century were those used exactly as Alexander Hamilton foresaw: to "restore the tranquility of the commonwealth" by a "well-timed offer of pardon to the . . . rebels." The pardon could bring rebels back into the fold, or it could repopulate the army by restoring deserters to service. President John Adams pardoned the rebels of Fries' Rebellion. President Thomas Jefferson pardoned deserters from the Continental Army. In order to fill up the army ranks to fight the War of 1812, President James Madison pardoned deserters and, after the war, pardoned Lafitte's pirates. President Andrew Jackson pardoned deserters. As the presidents well knew, pardons are a better signal than an armistice agreement to show that a war is truly over and that peace is restored.

A new situation called for presidential clemency during the Civil War. This was the first American war in which a substantial part of the army was conscripted. Desertion and draft evasion were huge problems. About 88,000 soldiers deserted during 1864 alone.[14] So many soldiers ran off to Canada that the border had to be closed to them. President Abraham Lincoln pardoned deserters on condition that they return to their units and fight on, and many did. He liberally pardoned supporters of the Confederacy in a textbook example of using pardon to undercut a rebellion. President Andrew Johnson pardoned the leaders of the Confederacy, and President Ulysses S. Grant also used pardons to clean up after the Civil War.[15] The Supreme Court may have decided in *Ex parte Wells* that a pardon "is an act of mercy flowing from the fountain of bounty and grace,"[16] but presidents used pardon very much the way they used their armies—to end wars.

The public pronouncements accompanying pardons made it clear that, even if the courts were not, the presidents were struggling with the contradictory considerations inherent in a pardon. Like any other human beings, the presidents were often moved to action by sadness and pity. President Lincoln was especially so:

> I hope there will be no persecutions, no bloody work after the war is over. No one need expect me to take part in hanging or killing these men, even the worst of them.

But, like the others, he understood the dangers of pardoning, dangers fully explained by Bentham, Kant, and Hegel. Lincoln recognized the problem of undermining the deterrent effect of punishment:

> Long experience has shown that armies cannot be maintained unless desertion shall be punished by the severe penalty of death.[17]

President Johnson also recognized that there was such a thing as too much kindness:

> The excellence of mercy and charity in a national trouble like ours ought not to be undervalued. Such feelings should be fondly cherished and studiously cultivated. When brought into action they should be generously but wisely indulged. Like all the great, necessary, and useful powers in nature or in government, harm may come of their improvident use, and perils which seem past may be renewed, and other and new dangers be precipitated. But by a too extended, thoughtless, or unwise kindness the man or the government may warm into life an adder that will requite that kindness by a fatal sting from a poisonous fang.[18]

President Johnson understood also the danger of undermining retributive justice:

> Some of the great leaders and offenders must be made to feel the extreme rigor of the law—not in a spirit of revenge, but to put the seal of infamy on their conduct.[19]

And he understood the danger of undercutting an individual's own decisions and standing as a human being:

> When men have offended against the law their appeal is for mercy, not for justice. In this country and under this Government violators of the law have offended against a law of their own making; out of their own mouths they are condemned—convicted by their own judgments—and, under a law of their own making, they cannot appear before the seat of mercy.[20]

Easing the aftermath of a war was not, of course, the only use of presidential pardoning power during the nineteenth century. The use of pardons gradually increased, and pardons were granted, as Hamilton anticipated, for a wide variety of reasons. Between 1860 and 1900, 49 percent of the applications for presidential pardons were granted. A more interesting statistic, the percentage of all federal prisoners pardoned (those who applied and those who did not), was not available until 1896. In that year, there averaged 64 acts of pardon for every 100 federal prisoners. In the next five years, the ratio between acts of clemency and prison population was, on average, 43 percent.[21]

That is a lot of pardons. But twentieth-century readers need to understand them from a nineteenth-century perspective. Most important, probation and parole procedures, while available in some states, would not be widely practiced until well into the twentieth century. 'Good conduct' laws—time off for good behavior—were unheard of. There was no Social Security, no Aid to Families With Dependent Children, nor any other social safety net, so the dependency of an imprisoned wage earner's family was truly serious business. The insanity plea was in its infancy. The important benefits of many twentieth-century procedural protections were unavailable. Federal offenses were far less serious and numerous than they are today. This was before the expansion of the criminal code, when the federal government enacted laws dealing with crimes the states could not handle alone, and when punishment for second offenders increased. As a result, federal offenders then were different in important respects from federal prisoners today. And since there were no federal prisons until 1895, federal offenders had to be housed in state institutions where the truly hardened criminals were held. So the need for pardons was significantly greater than it is today.

After 1885, the reasons for granting presidential pardons began to be recorded.[22] Some of them seem quaint and comfortable: the prisoner was "sincerely repentant," the pardon was granted in "recognition of upright living" or to "enable a farmer prisoner to save his crop." Some of the reasons are chilling and contemporary: "unsanitary and inadequate conditions in jail" and "health a menace to other prisoners." Some of the reasons have to do with the effort to treat the criminal fairly, including mitigating circumstances, doubt as to guilt, insanity, and—these are hair-raising—"mental infirmity of judge" and "dying confession of real murderer."

The majority of pardons, though, seem to have been granted by Presidents who were simply befriending the "sick and friendless." "Poverty" and "dependency of family" were often cited as reasons for pardons; so were youth, old age, infirmity, pregnancy, ill health, imminent death, and—perhaps most honest of all—"sympathy."

Would Kant, Hegel, and Bentham turn in their graves to learn that, almost exactly a century after their great tirades against pardons, as many federal offenders were pardoned each year in the United States as stayed in jail? That is a difficult question. I believe that it is more likely that they would be saddened that one hundred years had brought so little progress toward the humane and just legal systems that each one believed would eliminate the need for pardons.

5

Pardon and the Rehabilitative Ideal

Under the influence of Bentham's optimism, philosophers, social scientists, and jurists alike hoped that the newborn twentieth century would be an era of penal reform. The hope was that the humanity of law and the power of science would be harnessed together to achieve social goals; American prisons would become rehabilitation centers from which offenders would issue forth, transformed into productive members of society. Since sentences would be fitted to the rehabilitative needs of each individual, there would no longer be a need for the institution of pardon. Pardons would wither away, and they would not be missed—for their presuppositions of moral guilt were inconsistent with the new faith that offenders were sick or defective or confused, but never wicked.

As attorney William Smithers reported in the *Annals of the American Academy of Science* in 1910,

> The readjustment is going steadily forward and there is little doubt that in the near future the people of this great nation will at last fully emerge from the cloud of antiquity, discard the old cruel, useless and futile criminal laws and procedure and adopt sane, humane and rational measures for the protection of society from its defective members and provide for their proper care and reformation. Then every crime, when its perpetrator is discovered, will mean special study and treatment of the culprit according to enlightened methods and ends, *and the par-*

doning power will change from an active function into an interesting historical tradition.[1]

Smithers's prediction was half right. The percentage of federal offenders pardoned did drop in a jagged, steep curve between 1900 and 1936, to 2.7 percent,[2] as parole took over many of pardon's functions. During the same period, the Supreme Court, in *Biddle v. Perovich*, repudiated the concept of pardon as a private act of grace and frankly acknowledged that pardon was to be—like so many other institutions in those heady days—a tool for the public good. On the other hand, Smithers's prediction of a sane, humane, and rational penal system was mistaken, but that did not become clear for half a century.

The Rehabilitative Ideal

The rehabilitation theory of punishment that gained influence in the country between 1900 and 1960 was to have been the working out in practice of a theoretical union of the new behavioral sciences with the old utilitarian theories of Bentham. Bentham believed that punishment could serve the public welfare by operating in three different ways, by "incapacitation, reformation, and intimidation."[3] Of these, intimidation and the resulting deterrence were likely to do the most good, since they would affect the most people. Bentham went so far as to envisage perpetual imprisonment for some offenders in prisons painted black to "inspire salutary terrors" in all who came near.[4]

From Deterrence to Rehabilitation

Bentham's deterrence-through-intimidation justification for punishment presupposed that people—even criminals—are rationally calculating, pleasure-seeking beings. If people do not try to avoid pain, or if they are not able to foresee the consequences of their acts and alter their actions accordingly, deterrence will not work.

Those who wanted to put Bentham's theory of punishment to work in the American penal system faced two problems—one practical, one moral. The practical problem was that, as the new empirical social scientists turned their techniques on prisons, they could not find evidence that deterrence worked. Could Bentham have been mistaken about the rationality (or the hedonism) of criminals?

The second problem was this: While the imprisonment of an offender might serve society by setting an example, it did so at dreadful cost to the offender. Why should one person suffer behind black walls for the good of society? The deterrence theory, explained U.S. Attorney Julian P. Alexander,

> makes every punished criminal a martyr. The potential criminals in society have no right to expect a vicarious atonement by him who is in the toils. This idea was carried to a similar extreme in the request made by a modern German mother to the keeper of her boy. "If Fritz is bad," she wrote, "you need not whip him. Spank the boy next to him and it will scare him to death."[5]

Even this might be tolerable, for the sake of a greater good, if prisoners were congratulated for their contributions to society. But the opposite is true. Together with the loss of freedom comes a different hurt, "a single, monotonous accusation: 'You are a bad man and we hate you.' [Prisoners are] slapped in the face by society."[6] If a person is in prison to deter others from crime, then surely it is unnecessary cruelty to continue with the deceit that he is in prison because he is wicked.

How could a Utilitarian preserve the justification for punishment based on the expected good to society yet keep from using prisoners as means to societal ends? There was a way: The rehabilitation justification for punishment solved Utilitarianism's problems in a single stroke. Deterrence does not work?—then let reformation justify punishment. Deterrence sacrifices the happiness of one for the happiness of many?—then help the prisoners also, by remaking them into law-abiding citizens. This was Utilitarianism with a modern, scientific face. It kept the same goal—the increased happiness of society through reduced crime—but used a different means—the reform of criminals, not the deterrence of crime.

A powerful combination of forces fueled the rehabilitation movement. It was fanned by a new optimism, the faith that science could find a solution to problems that stumped the philosophers and that government, by the right combination of laws, could be a positive force for the public good.

Philosophical and Scientific Foundations

The rehabilitation movement did not share Bentham's presuppositions about the nature of humankind. Borrowing ideas from nineteenth-century German psychologists, from behaviorists, and ultimately from psychiatrists, the rehabilitation movement presupposed that humans are not the rational, calculating beings so dear to Bentham. Nor are they the jurists' "reasonable men." They are objects moved by causal forces, conscious and subconscious, psychological and social. This being so, it should be possible to isolate the causes of crime, eliminate them, and thus eliminate crime—the scientific solution.

Some looked for the causes of crime in society itself. John Dewey argued that "we are all in a social partnership in making crime."[7] In pretrial detention in Soledad Prison a half century later, armed robbery suspect George Jackson agreed. Shortly before he was either murdered by guards or killed in an escape attempt, he wrote

> I didn't create this impasse. . . . Did I colonize, kidnap, make war on myself, destroy my own institutions, enslave myself, use myself, and neglect myself, steal my identity and then, being reduced to nothing, invent a competitive economy knowing that I cannot compete? . . . It was a fool who created this monster, one unaccustomed to power and its use, a foolish man grown heady with power and made drunk, dizzy drunk from the hot air that inflates his ego. I am his victim, born innocent, a *total* product of my surroundings.[8]

If society creates crime, then a cure for crime must come from steps that would cure societal ills and injustices. In 1910, sociologist Charles Ellwood made it sound easy: "To eradicate crime and the criminal from society only requires that man shall attain to the same mastery over the social environment which he has already practically attained over the physical environment."[9]

Some causes of crime were thought to be psychological: brain disorders, diseases of the mind, and subconscious urges. In the early stages of the rehabilitation movement, the connections between crime and disease, penology and medicine, were analogical. A person who committed a crime was *like* a person with a disease: not to be blamed, certainly, but in need of isolation and treatment. As Alexander wrote,

> The modern Samaritan, as he goes about making clean the tenement, combating disease, and lifting up the fallen, begins to look with tenderness and sane sympathy upon those stricken with vice . . . as with a plague.[10]

In a 1955 criminology textbook, *New Horizons in Criminology* (the title itself is a sign of the times), the analogy was carefully drawn:

It is obviously both futile and unjust to punish him as if he could go straight and had deliberately chosen to do otherwise. It would be as foolish to punish him for having contracted tuberculosis.[11]

Gradually, with the growth of psychiatry, the analogy between crime and disease dropped out. Crime came to be considered a symptom of disease—not metaphorical disease but real, physical disease. As early as 1920, Alexander made an armchair diagnosis of the disease:

> The criminal is a man of diseased mind or morals. His malady is a real one and demands thoughtful and positive treatment. Experience proves that nearly 90 percent of all criminals are sick or diseased in body and a large proportion of crimes are attributable to physical causes.[12]

The humane response to sickness is sympathy and curative efforts; likewise, the humane response to crime is to put offenders in places of safety, 'correctional facilities,' and to carry out courses of treatment best designed to reform them. The people best able to determine the courses of treatment are not judges or juries, and certainly not legislators, but rather those closest to the prisoners— the prison officials and parole boards. According to Richard Singer, "The diagnosis and treatment of the criminal is a highly technical medical and sociological problem for which the lawyer is rarely any better fitted than a real estate agent or a plumber."[13]

Thus the primary tool of rehabilitation became the indefinite sentence. Instead of the mere lapse of time, sentences were measured by time until reformation, as demonstrated to the satisfaction of the appropriate officials. The 1972 Model Sentencing Act put it succinctly: "[O]ffenders shall be identified, segregated, and correctively treated in custody *for as long terms as needed*."[14]

It was clear to the utilitarian penologists that if treatment is to take the place of punishment, a variety of philosophical niceties are no longer needed and may be allowed to waste away. Equitable sentences, punishment that fits the crime, guilt, responsibility—all these concepts, they argued, are out of date, belonging to outmoded philosophies of the past. And justice? "The very word *justice* irritates scientists," wrote Karl Menninger, a psychiatrist and author of *The Crime of Punishment*, the unintentional epilogue to the rehabilitation movement. "No surgeon expects to be asked if an operation for cancer is just or not. . . . Behavioral scientists regard it as equally absurd to invoke the question of justice in deciding what to do with a woman who cannot resist her propensity to shoplift. . . . "[15]

So-called 'obituaries' for retributive justice appeared in print. The 1966 *Manual of Correctional Standards* stated that

> Punishment as retribution belongs to a penal philosophy that is archaic and discredited by history.[16]

Henry Weihofen may have offended professors, but he expressed the predominant view of criminologists when he wrote,

> all of this abstract philosophy about punishment as a requital for crime has a musty smell about it, the smell of a professor's study. It does not breathe the air of reality. . . . The modern behavioral sciences have shown that arm-chair abstractions about the "justice" of retribution by philosophers who reject human experience [footnote: Immanuel Kant, *Philosophy of Law*] are sadly defective in human understanding, not to say human sympathy.[17]

To fully appreciate how much criminology has changed in principle, and the philosophical significance of the change, one should study the dogma that greeted the criminology students of the 1950s. A table on the inside cover of a textbook summed it up:[18]

	1800	1955
Stage of correctional development	Varied punishments for varied responsibility	Psychotherapy
Significant influence or events	French Revolution and rise of middle class	Camps and probation, in-service diagnostic clinics
Theory of crime causation	Free choice of evil	Personality frustration
Treatment used	Penal colonies, solitary confinement	Group therapy, counseling, small family units
Purpose of treatment	Rehabilitation, with penitence	Personality adjustment, social orientation
School of criminology	Classical	Multiple causation
Major control groups	"The people," politicians	Professionally trained clinicians

Source: Copyright ©1955 by H. Barnes and N.K. Teeters. Reprinted with permission.

If that is what was to be made of punishment, what was to be made of pardon?

The Reduced Role of Pardons in the Twentieth Century

One of the expectations of the rehabilitation ideal was that the penal system would become so much improved that pardons would no longer be necessary. Looking back, it is now clear that reality did not quite live up to the ideal. But pardon's role was significantly reduced, both in theory and in practice.

The Expected Role of Pardons in a Rehabilitative System

The rehabilitative model was to have taken away many of the traditional uses of pardon in two ways. First, many of the needs that pardons met were now to be met by the indeterminate sentence and the parole board. In theory, these options individualized sentences to fit the needs of the offender, so pardons were not needed to alleviate the "sanguinary and cruel" individual effects of a criminal law written in broad generalizations. If criminals saw the error of their ways, or if particular circumstances made a penalty unnecessarily harsh, the sentencing judge or the parole board would presumably make the suitable adjustments within legislated boundaries.

Second, the philosophical presuppositions of the rehabilitative model undercut many of the former justifications for pardon. For example, *equity* had often been cited as a reason for pardon: Pardons were granted if two equal partners in crime had received unequal sentences, or a hapless accomplice received a harsher sentence than the brains behind a crime. But equity was no longer to be a value in sentencing. Sentences were measured by the future needs of the offender, not the circumstances of other offenders.

Reduced culpability had at one time been a reason for pardon. A feeble-minded or youthful offender caught in the law enforcement net had a case for pardon under the principle that those who cannot be blamed cannot be punished. But, in the brave new world, 'blameworthiness' was a relic of an older, more vengeful system preoccupied with wickedness. If punishment is benevolent treatment, it can be extended to all those who are in need of reform. Indeed, the rehabilitation model presented no theoretical block to treating/punishing weak-willed people before any crime was committed.

Proportionality also had been a justification for pardon. Pardons could prevent the injustice of a punishment more serious than the crime and could factor into the equation the degree to which offenders had already suffered from the natural consequences of their crimes. But, again, proportionality was not a value in the rehabilitative system.

And, finally, *pity* once prompted many pardons. But why should people be pitied when they are receiving help that will make them well-adjusted citizens? Here is the argument by analogy: One feels pity for a person whose leg movement is restricted by a plaster cast, but the pity is occasioned by the broken bones, not by the presence of the cast. Certainly, pity does not justify removing the cumbersome cast. Just so, one may pity a person who needs treatment/punishment, but not because he is getting what he needs.

The Actual Role of Pardon

So this is what is *supposed* to have happened: Punishments were to be replaced by effective treatment programs; pardons were to be replaced by individualized, indeterminate sentences and parole. But reality did not quite live up to expectations. Most states did move ahead with indeterminate sentencing. Most states set up parole boards that determined whether an offender would get out of jail. Many of the justice-based philosophical concepts were blurred, forgotten, or confused. But rehabilitating offenders turned out to be far more difficult than anyone expected. Moreover, the actual changes made in the system did not match the reform rhetoric. The sign over the prison gate changed, the titles of the wardens changed, but—for the most part—changes inside the prisons were not as radical as the reformers had hoped. Thus, the need for pardon lessened, but it did not disappear.

There were philosophical reasons, as well, to explain why pardons did not disappear. As much as the movement's theoreticians may have accepted the idea that considerations of justice are irrelevent, the general population did not. F. H. Bradley pointed out that the "vulgar" view of punishment was still based on principles of retributive justice,[19] and polls conducted by philosophers (of all people) bore him out.[20]

Because many of those in a position to grant pardons generally shared the vulgar view, pardons continued. In 1936, for example, the federal prison population was 14,000, and 2,100 were released on parole. Another 1,500 applied for pardons, and President Franklin Roosevelt pardoned 390 of them.[21]

Some presidents—Theodore Roosevelt, Calvin Coolidge, Franklin Roosevelt, Harry Truman—used pardons in time-honored ways, to release those imprisoned during the heat of war, for example.[22] Also, several notorious or extraordinary actions took place during this time (1900 to 1960). In a Fourth of July proclamation in 1902, Teddy Roosevelt pardoned Filipinos who committed crimes during the time their country was ruled by Spain. At Christmas time of 1921, Warren Harding freed socialist and presidential candidate Eugene Debs, imprisoned for violating the Sedition Act. Herbert Hoover was reportedly an easy mark for a sad story, pardoning more than most presidents did.

This was the time, too, of scandalous pardons and pardons for sale. In Texas, for example, Governor James Ferguson pardoned 1,774 inmates before he was impeached for so flagrantly abusing the pardoning power. Succeeding him, Governor W. P. Hobby and Governor Miriam Ferguson pardoned 1,703 people between them.[23]

Changes in the Philosophical Concept of Pardon

While the institution of pardon was being carried on, the concept of pardon as a private act of grace was in serious peril. Whether measured by retributive or utilitarian standards, there was understood to be a proper punishment for each case. Using the pardoning power to reduce or change that punishment was *ipso facto* improper or at least in need of justification. Grace—the personal inclination to show pity—was not a justification. Governor John Jay of New York had said it long before, "To pardon or not to pardon does not depend on my will, but on my judgment; and for the impartial and discreet exercise of this authority I am and ought to be highly responsible."[24]

Two philosophers, both writing in 1910, argued analogically that pardons could not be gifts, offered or withheld at the will of the president. The first, Charles Bonaparte, argued that there was no difference between taking a bribe and giving a pardon for reasons of mercy alone. The acts are similar in that each involves a decision based on nothing more than self-satisfaction. Just as taking a bribe is unconscionable, granting unjustified gifts of grace is a violation of the public trust.[25]

The second, William Smithers, suggested an analogy between

the presidential power to pardon and the presidential power to call out the militia. Whether the militia is called out is a practical, hard-headed decision on the President's part. Calling out the militia when it is not needed is dangerous. "But he can no more honestly withhold a pardon in a proper case than he can refuse to call out the militia when the preservation of public peace demands it."[26]

Both analogies suggest that pardon is not so much a gift freely given but a decision carefully made on the basis of considerations of strategy and justice. What is remarkable about the arguments is that they both imply that the benefit of the pardon to the recipient is of only incidental importance. Neither benevolence nor beneficence is in any way essentially connected with granting a pardon. If this is so, then the idea that pardon is an act of grace makes no sense at all.

Finally the Supreme Court too gave up on the doctrine of *U.S. v. Wilson*, which had defined pardon as an act of grace.[27] The first crack in the doctrine appeared in *Burdick v. U.S.* Burdick, a newspaper reporter, used his right against self-incrimination to avoid testifying about a customs fraud in New York. President Coolidge offered him a full pardon, which would have had the effect of forcing him to testify. Burdick refused to accept the pardon, and the court supported him, saying, "the grace of a pardon may be only a pretense."[28] The decision was tangled in inconsistencies, but it marked a weakening in the doctrine of pardon as grace.

Coolidge took matters into his own hands in 1923 and insisted on his right to pardon even without the consent of the offender. When a man named Craig, bent on martyrdom, refused a pardon, Coolidge ordered that he be expelled from jail and the doors locked behind him. Coolidge was ready to drop even the pretense of the grace of a pardon.

Four years later, the Supreme Court redefined 'pardon' by ruling that a pardon does not have to be accepted to be valid. The case was *Biddle v. Perovich*. Vuco Perovich was serving a life sentence, his death sentence having been commuted by President William Taft some twenty years earlier. But Perovich had changed his mind and wished to be executed. He argued that he had never been pardoned, since he had never accepted the commutation. The court rejected his argument. Justice Oliver Wendell Holmes spoke for the court: "A pardon in our days is not a private act of grace from an individual happening to possess power."[29]

In this way, the Supreme Court abandoned the concept of pardon enunciated in *U.S. v. Wilson*. By the same decision, they renounced the concept of pardon inherited from English common law. "[Pardon] is part of the constitutional scheme," Justice Holmes wrote. "When granted, it is the determination of the ultimate authorities that the public welfare will be better served by inflicting less than what the judgment fixed."[30]

Like punishment, pardon was to be a servant not of justice, not of the sovereign's benevolence, but of the public good.

6

The Retributivist Backlash

How pardons are understood and how they are used depends, to a large extent, on the values and procedures inherent in a system of punishment. The 1970s in the United States saw an important change in the predominant theory of punishment—the decline of the rehabilitationist model and its replacement by a view of punishment that more closely approximates retributivist ideals. That change set the stage for a new view of the role pardons should play in the U.S. system of justice.

The Decline of the Rehabilitative Ideal

As it turned out, the reports of the death of retributivism were, to paraphrase Mark Twain, exaggerated. After dominating penal systems for a century, the medical rehabilitation model of punishment began to weaken. No doubt its decline was aided by scientists, by moral philosophers, and especially by prison inmates. But the most serious wounds were self-inflicted. For a primary characteristic of the rehabilitation model turned out to be hypocrisy—the benevolent-doctor metaphor that made considerations of justice irrelevant—and the inability to live up to its lofty goals, a failure that made justice considerations all-important again. George Bernard Shaw summarized the problem: "reformation is a false excuse for wickedness."[1]

Ready to fill the void was a new retributivism. It provided the theoretical underpinnings of a system of punishment based on 'just deserts,' in which judicial discretion is sharply reduced and the measure of punishment is the seriousness of the crime. Supported by liberals, who saw it as a way to restore justice to what had become a horrible system of megaprisons, and supported by law-and-order conservatives who were tired of 'coddling' criminals, the retributivist view of punishment became, with surprising speed, the basis of a reform movement of its own.

The Practical Failures of Rehabilitation

In keeping with Bentham's metaphor that punishment is capital hazarded in the expectation of profit, it was appropriate for people to inquire what American had gained for its investment in rehabilitation. A variety of statistical research conducted between 1950 and 1970 pointed to the same conclusion: very little gain could be statistically verified. In 1975, a review of 289 studies of rehabilitation programs came up empty; no evidence could be found that any therapies reduced recidivism—or that they did not.[2] The National Academy of Sciences also concluded, after reviewing hundreds of studies on rehabilitation, that the claims made for these programs had been neither proved nor disproved.[3] The only factors seeming to correlate with the institution of the new procedures were a rising crime rate, rising costs, and—significantly—longer sentences. The abundance of confounding factors made it nearly impossible to draw any firm conclusions from these statistical correlations, but evidently the millennium had not yet arrived.

Injustices in the Corrections System

The disappointing lack of results might have been tolerable (Americans being accustomed to getting nothing for their tax dollars) but for increasing signs that rehabilitation models were imposing costs— most importantly intangible damage to standards of justice that people believed in. As professor and attorney Marvin Zalman hypothesized, "the sudden decline of the rehabilitation assumptions stems not only from the growing awareness of the limits of rehabilitation, but also from a renaissance of concern for justice."[4]

The most strident complaints of injustice came from the prison inmates themselves, a usually mute group given voice by two unusual events. The first was the 1971 prison riot at Attica in New York. The second resulted from protests against the Vietnam War, when large numbers of middle-class, well-educated young people entered the prisons. Although they did not usually stay long, they heard and saw enough to know that the rehabilitation model was not working justly.

They heard two main complaints. One was of the dehumanizing effect of a system that kept people in jail until they, either hypocritically or actually, accepted that they were sick and maladjusted and made demonstrable efforts to fit into the 'healthy' society that had put them there. The other was of the comparative injustices that resulted from a system with so much discretion that the length of a prison sentence was unrelated either to the crime committed or to the sentences of others who had committed similar crimes. It was a classic double-bind: If rehabilitation did work, it was an intrusion on an important liberty—the liberty to create oneself. If it did not work, it resulted in an unjustified and unjustly assigned prolongation of pain.

The Dehumanizing Effects of the Rehabilitation Model

A primary cause of the discontent that exploded into the Attica prison rebellion was the seemingly arbitrary and dehumanizing operation of the parole system.[5] Once inside, it was difficult for prisoners to predict when they would get out. Prisoners were denied even the comic-book solace of counting off days with hash marks on the concrete wall. Sentences ended when the parole board was convinced that the prisoner was no longer dangerous, a criterion that added a course in acting to the usual crime-school curriculum. What evidence aided the boards' decisions?—in some cases, only a seventeen-minute interview with two parole officials present, one to talk to the inmate, the other to read the record.[6]

The criteria for gaining parole were unclear and seemingly arbitrary. The 1971 case of *Monks v. New Jersey State Parole Board* is illustrative. William Monks was serving an indeterminate sentence for a murder he had committed when he was a child. Over and over again, his applications for parole were denied and no reasons were given. Please tell me, he begged the parole board, "what [is] necessary to convince the board that [I am] a good parole risk," so that I can "be in a position to behave in any way the Board expected."[7] The parole board answered that giving reasons for their decisions was against their policy.

In fact, inmates had the best chance for parole if they displayed good behavior and if they pronounced themselves profoundly repentant. (It also helped in some cases to be willing to confess to other crimes, allowing police to clear up their unsolved crimes files.) This much parole discretion prompted author and felon George Jackson to write that there were only two ways to leave San Quentin—in a meat wagon or, as he said, "licking at the pig's feet." It was not possible, he said, to leave standing up.[8]

C. S. Lewis had foreseen the dehumanizing effects long before:

> To be taken without consent from my home and friends; to lose my liberty, to undergo all these assaults on my personality which modern psychotherapy knows how to deliver; to be remade after some pattern of "normality" hatched in a Viennese laboratory to which I never professed allegiance; to know this process will never end until either my captors have succeeded or I have grown wise enough to cheat them with apparent success—who cares whether this is called punishment or not? That it includes most of the elements for which any punishment is feared—shame, exile, bondage, and years eaten by the locust—is obvious.[9]

Prison inmates understood what Lewis meant. "Society may have a right to punish [offenders]," an inmate said, "but not a hunting license to remold them in its own sick image."

What made this violation of integrity important was that it violated the premise that humans are rational beings whose acts—even criminal acts—result from reasoned decisions. The 'pathological' label on antisocial behavior was sold as a scientific judgment, but it may also have been political. Its effect (if not its intent) was to reduce dissent—discrediting society's critics by labeling them 'sick,' subjecting them to painful 'treatment,' and, by focusing public attention on the criminal's disorders, distracting attention from the disorder and injustice in society. It took the civil disobedience tactics of civil rights and antiwar demonstrators to finally damage the presupposition of the rehabilitationists, that no one *rationally* chooses to break the law.

The Injustice of the Indeterminate Sentence

A second major line of criticism focused on the practice of indeterminate sentencing itself, including the excessive length of sentences and the comparative injustice of the punishments.

Although nothing in theory made indeterminate sentences longer than any others, in practice indeterminate sentencing substantially increased the time offenders spent in prison. Harvard law professor Alan Dershowitz's 1974 study of indeterminate confinement drew this conclusion:

> It is clear that no matter what other factors govern sentence length, the inevitable result of the indeterminate sentence is that sentences of over five years will predominate, and in a definite sentence state sentences of under five years will predominate.[10]

In rehabilitationist theory, the crime is the occasion—but not the measure—of punishment. The crime serves only to attract the attention of the state, so the relative harmlessness of a criminal act sets no natural limit on the punishment it calls forth. As one might expect, horror stories abound. John Lynch, the subject of the California Supreme Court decision *In re Lynch*,[11] was a person with "superior intellect" and a 1958 conviction for indecent exposure. In 1967, he was discovered masturbating in his car in the darkened parking lot of a drive-in restaurant, with no one around but an apparently shocked policeman. Lynch was given an indeterminate sentence. By the time he appealed, he had served five years, three and a half of those in maximum security in Folsom prison. He was fortunate that the California Supreme Court had ruled that punishment is cruel and unusual if its duration is excessive in relation to the offense. Otherwise, he might be in prison yet, indecent exposure being an especially stubborn obsession.

In the treatment of Lynch and other offenders caught under repeat offender or dangerous offender programs, prison officials blurred the distinction between indeterminate punishment and preventive detention. Having been identified as particularly prone to crime, these people were warehoused in jail not so that they would be punished, but so that they would not commit the same crime again.

The issue of proportionality between crime and punishment was raised in *People v. Levy*.[12] William Levy complained that, although he had been convicted of a misdemeanor only, he was serving an indeterminate sentence in San Quentin. The California Court of Appeals, unencumbered by old-fashioned worries about justice, replied this way:

> The emphasis that appellant places on the fact that he was originally convicted of a misdemeanor, and now finds himself in San Quen-

tin, possibly for life, is misplaced. The argument would be sound only were his confinement punishment. . . . [But] the purpose of confinement is to protect society. . . . [13]

This was one of the results of disassociating the seriousness of the crime and the seriousness of the punishment.

Another result was that two people who committed the same crime under similar circumstances found themselves serving substantially different sentences when one seemed less likely than the other to reform. Rehabilitation justified individualized sentences, individualized sentences justified discretion, and discretion resulted in enormous disparities between sentences for similar crimes. Describing California, the state with probably the greatest sentencing discretion, University of Chicago law professor Franklin Zimring wrote,

> Felons . . . could receive anything between probation and life imprisonment for the same crimes. Persons convicted of similar crimes received grossly different punishments. Those committed to prison arrived at the gate not knowing whether they would stay for one year or for ten. [14]

Most interesting, state indeterminate sentencing laws were sometimes set up in ways that discriminated against women. Since women were assumed to be more likely to be rehabilitated, early indeterminate sentencing laws applied only to them, the states having less hope for reforming their male offenders. [15] A New Jersey law set *higher* sentences for women than for men, reasoning that women— being pliable—should be given ample opportunity for reform. [16] The courts eventually struck these practices down as violations of the equal protection clause.

Courts were not as effective in responding to the comparative injustices suffered by minorities, by the poor, and by those who committed 'real' crimes (as opposed to white-collar crimes), many of whom ended up spending more time in jail than their richer, whiter counterparts. [17] The best way to get parole was to be the general sort of person who does not frighten the general sorts of people on parole boards—that is, to be a white, white-collar criminal. [18]

Thus, the indeterminate sentence, the work horse of rehabilitation, turned out to be a Trojan horse. It introduced intolerable opportunities for arbitrariness, uncertainty, and injustice into the system of corrections. It had to go.

All these considerations came together in the American Friends

Service Committee report on crime and justice in America, which was published in 1971 under the title, *The Struggle for Justice*.[19] Quoting extensively from participants in the Attica prison rebellion, it chronicled the failure of the rehabilitative model and called for the abolition of indeterminate sentences.

For philosophers, these events accomplished what even Kant had not entirely succeeded in doing—making brutally clear the dangers of having a system of punishment in which justice was irrelevant. One by one, and then in increasing numbers, the voices of philosophers were heard across the land, advocating a return to the principles of retributive justice. The titles of their scholarly articles tell the tale: "The Retributivist Hits Back."[20] "A Plea for Deserts."[21] "Why Do We Punish: The Case for Retributive Justice."[22] "The Restoration of Retributivism."[23] "Renaissance of Retributivism."[24]

The Restoration of Retributivism

Although philosophers had been suggesting for many years that the wholesale rejection of retributivist principles should be reexamined,[25] it was not until the 1960s that the job began in earnest.

The Early Stages

The reexamination began with some philosophers who thought that both retributive and utilitarian principles had important roles to play in a theory of punishment. Even while arguing that retributivism was dead and in the advanced stages of decay,[26] H. L. A. Hart suggested a compromise, an "integrated rationale." A theory of punishment, he argued, requires two principles. The first is a justifying principle, a General Justifying Aim: what justifies the institution of punishment *as a whole*? Hart's answer was close to the utilitarian line: society punishes in order to deter crime or reform criminals—to protect people from crime. But the question of the distribution of particular punishments is a different question. Only considerations of retributive justice—what the offender deserves—can justify particular punishments. Because offenders' fates should depend on their choices, only responsible acts can be punished.

From the 1960s into the 1970s, philosophers tried to sort out the conflicting or complementary demands of utilitarianism and retributivism. They wanted to pick and chose the best from both theories, to come up with a multivalued theory of punishment that acknowledged that both the general welfare and justice are to be taken into account. The resulting compromise efforts[27] worked with different hierarchies of obligations. Although it is dangerous to generalize about as heterodox and cantankerous a group as philosophers of law, one may venture to identify some trends in the early days of the restoration of retributivism.

At first, retributive justice was given a negative role, as a set of principles limiting morally possible punishments. Utilitarian considerations of deterrence, rehabilitation, and the good of society were to be used to identify all the cases in which punishment would be good. Then, 'negative retributivism' (a sort of retributivist morality police) would come through and knock out the punishments that could not be justly imposed. It was unjust and impermissible to punish people who had not done wrong, either because they had not committed a crime or because they could not be blamed for the crimes they had committed. And it was unjust and impermissible to impose punishments of a severity disproportionate to the seriousness of the offense. In this way, retributivism was limited; it was resurrected like a zombie to do a particular task—to serve as a check on utilitarianism, which justified too much.

It was not long before philosophers began to reconsider the limitations they had imposed on retributivism, arguing that the same retributive considerations that limit punishment can justify punishment. Considerations of justice may not only be necessary, but they may also sometimes be sufficient conditions for punishment.

Philosopher Joseph Weiler argued that a "crucial aim of punishment is to restore the equilibrium of benefits and burdens that the criminal law is designed to define and maintain."[28] The aim of punishment is not just to make a happier society, although that is important. Punishment is also supposed to help make a just society. A just society is an end in itself. So retribution is *a* justifying aim of punishment.

Sidney Gendin also gave retributive considerations an expanded role. He offered "A Plausible Theory of Retribution"[29]: A *prima facie* reason for punishing a person is that he is morally responsible for a crime he has committed. If, however, a punishment has harmful consequences that outweigh its good consequences, it ought not to be imposed. However, it does not follow that a punishment has to have beneficial consequences to be justifiable. "Where all we can do is speculate about the utility, the fact of guilt is decisive."[30]

John Finnis opened the crack a little wider, arguing that there is a *prima facie* obligation (not just permission) to punish, based on retributive considerations. He began with Kantian ideas: It violates an obligation to law-abiding citizens when criminals are not punished.[31] But this is only part of the story. It more seriously violates the rights of law-abiding citizens "to punish criminals when it is clear that the punishment will lead to more crime, more unjust acts by criminals and more danger and disadvantage to law-abiding citizens."[32] Thus, while the retributive restoration of justice is the most specific and essential aim of punishment, it should not be pursued regardless of cost.

These tentative steps toward rehabilitating retributivism and making it again a productive part of society were for a time confined to the philosophical community. But retributivism, under the name of 'commensurate deserts' became a rallying point for reformers of all kinds, with the 1976 publication of *Doing Justice*.

A New Retributivism

Doing Justice was the report of the Committee for the Study of Incarceration.[33] Created and funded by two liberal foundations,[34] the committee included liberal thinkers of some distinction—among others, Senator Charles Goodell, a former U.S. Senator; Alan Dershowitz, a Harvard law professor; Eleanor Holmes Norton, the chair of the New York City Commission on Human Rights; and Andrew von Hirsh, a Rutgers professor of criminal justice and author of the report. The report blasted the rehabilitative ideal, arguing that what offenders deserve—and that alone—justifies their punishments and determines how severe they should be. Von Hirsch used phrases and ideas that this nation had scarcely heard for two hundred years:

> Certain things are simply wrong and ought to be punished. This we do believe.[35]

For a justification for punishment, von Hirsch turned to one interpretation[36] of the ideas of Kant as "a useful place to begin."[37] Von Hirsch paraphrased the ideas this way:

> To realize their own freedom, ... members of society have the reciprocal obligation to limit their behavior so as not to interfere with

the freedom of others. When someone infringes another's rights, he gains an unfair advantage over all others in society—since he has failed to constrain his own behavior while benefiting from other persons' forbearance from interfering with his rights. The punishment—by imposing a counterbalancing disadvantage on the violator—restores the equilibrium: after having undergone the punishment, the violator ceases to be at an advantage over his non-violating fellows.[38]

If punishment can be said to have a purpose, it is to achieve or restore justice. And justice needs no justification.

Von Hirsch added several other philosophical presuppositions.[39] First, the liberty of each individual is a primary good, to be protected. Second, the state should use the least restrictive alternatives, interfering as little as possible in an offender's life. And third, the requirements of justice should constrain not just the trial but also the postconviction treatment of the offender.

These principles put limits on the form that justifiable punishment can take. Punishment must be closely related to what the offenders have *done*, not to what they *may do*. The punishment, in other words, must be commensurate with what the offender deserves, and what the offender deserves depends on the offense committed. The more serious the crime, the more serious the punishment. This is the principle of "commensurate deserts."

In this point of view, the seriousness of a crime is measured by the amount of harm generally done by acts of that sort and by the degree of culpability. The seriousness of punishment is measured by the general tolerance for suffering in the society.

The Recommendations of Doing Justice

Given this view of punishment, *Doing Justice* concluded that many of the then current sentencing practices were unjust and unjustifiable. Von Hirsch made a series of recommendations for reform:

First, they recommended less punishment overall. Sentences for first offenders would be sharply reduced, with very few sentences exceeding three years. The use of incarceration would be confined to very serious crimes, and alternatives such as warnings, fines, and deprivations of leisure time would be used for all other offenses.

Second, they recommended less disparity and discretion. Offenders with similar offenses would receive similar sentences. The wide judicial discretion would be sharply narrowed by sentencing guidelines and presumptive sentences, with some limited variation for mitigating and aggravating circumstances. There would be no early release.

Von Hirsch envisioned a sentencing system in which every person who commits a crime knows in advance what is likely to happen as a result; in which every person serving a sentence knows exactly when it will end; in which every person knows that the worse the offense is, the worse the punishment will be; in which an offender can expect to be treated just like any other who has committed the same crime; and in which an offender can expect to be treated humanely. What the committee envisioned, in short, was Beccaria's vision—a state where laws are humane and punishments inexorable.

Such a sentencing system might deter crime (the data were mixed but vaguely promising), but its real selling point was that it might achieve justice.

And sell, it did. People who had been dissatisfied (or outraged) by the criminal justice system found *Doing Justice* to be The Answer. The Twentieth Century Fund's study, *Fair and Certain Punishment*,[40] published soon after *Doing Justice*, provided a detailed approach to sentencing based on desert. Legislatures and parole boards moved ahead with "incredible alacrity"[41] to put the reforms into effect. While it must be pointed out that the 'less discretion' recommendation had more fans than the 'less punishment' recommendation, legislatures moved quickly to enact presumptive sentencing legislation and, in some states, to abolish parole altogether.

Nobody has yet written an article entitled, "The Revenge of the Retributivists," but it should be written. The 1970s marked at least a temporary triumph of retributivist theory over the rehabilitation ideal that had dominated penal theory for more than a century.

Problems in Doing Justice

The new retributivism underlying *Doing Justice* was not without conceptual weaknesses and vulnerabilities. Of these, two especially need to be pointed out.

First, a contradiction exists between the means chosen and the ends identified in *Doing Justice*. The new-retributivist ideal was to sentence each individual to a punishment that the individual deserves, as measured by past conduct. The specific aspects of past conduct that determine the punishment are the seriousness of the harm done by the offense and the culpability of the offender. Offenses are in-dividual acts, and—until they are placed in legal categories—they are unique. No two crimes (even those with the same legal description) cause just the same harm; no two criminals (even those convicted of the same crime) have just the same degree of culpability. So sentences must also be unique. For this reason, 'just desert' is individualized desert.

The problem is that *Doing Justice* argued that the means for assigning 'just deserts' is the presumptive sentence, which categorizes past offenses by the amount of harm *generally* done by acts of that sort and by the degree of culpability *usually* associated with acts of that sort. The new retributivists wanted to abolish discretion and individualized sentences and, *at the same time*, to give all offenders what they individually deserve. If they were really serious about desert-based sentences, the new retributivists would have had to in-crease rather than decrease discretion.

This von Hirsch was unwilling to do, recognizing that discretion in a retributivist system has the same practical and moral evils as discretion in a rehabilitative system. And getting rid of those evils was what *Doing Justice* was all about. Von Hirsch wisely chose the occasional injustice brought about by broad rules over the systematic injustices of the rehabilitation ideal; still, it was a trade-off.

The second problem has to do with the concept of 'desert.' What facts about offenders make them deserve punishment? That depends on the justification of punishment, about which there is disagreement even among retributivists. If punishment is primarily justified as the requital of evil for evil, then the offender's moral character is the central issue. This is 'moral desert.' If punishment is primarily justified as forcing a person to give back an advantage unjustly attained, then the harm done is the central issue. This is 'legal desert.' The *Doing Justice* model attempted to use both criteria, basing a judgment of the seriousness of a crime on both culpability and harm.

Moral desert and legal desert are undeniably closely related, in theory and in practice. One who does harm usually acts immorally. And one who acts immorally usually does harm. However, there will always be cases when the punishment that is deserved based on moral desert is different from that based on legal desert. *Doing Justice* put primary emphasis on legal desert. As a result, there will always be cases in which the *Doing Justice* model imposes punishments deserved in one sense but undeserved in another. This might occur in cases dealing with unjust laws, honorable motives, or unjustly disadvantaged offenders, for example. Again, the new retributivism substituted a lesser form of injustice for a greater. That this is progress cannot be denied, but there is still work to be done. The institution of pardon may be the best for the job.

7

Pardoning in Transition

To this day, courts, legislatures, and philosophers are still working out the details of a system of punishment based on commensurate deserts. Pardon and punishment have always been closely connected, both in theory and in practice. So it is no surprise that, as the system of punishment disentangles itself from the old theory—even while carrying the weight of the new theory—the pardoning power, stumbling beside punishment, needs to redefine itself and *its* new direction. To date, there has not been a major change in the theoretical justification for pardon that parallels the shift in justification for punishment.

Major Pardons of the 1970s

The unusual features of President Ford's pardon of Richard Nixon suggested to some people that a major break with traditional uses of pardon had already taken place. It had not. Although it is unusual for a pardon to precede a trial, and even more unusual for a pardon to be granted before charges are brought, there is precedent for both. In 1867, Andrew Johnson granted a pardon to a man named A. H. Garland for any offenses he may have committed while supporting the Confederacy during the Civil War. The U.S. Supreme Court wrote (and even the dissent conceded) that

> [The pardoning power] extends to every offense known to the law, and may be exercised at any time after its commission, either before legal proceedings are taken, or during their pendency, or after conviction and judgment.[1]

Moreover, amnesties, a form of pardon, are almost always granted to classes of people who have not been charged with any crime. In fact, the only unusual aspect of the Nixon pardon was its notoriety, and even that diminishes in comparison to the pardons granted during the 'Great Age of Pardons' before the Enlightenment.

The Nixon pardon was very much in the tradition of *Biddle v. Perovich*—a President using a pardon as a tool for the public good, as he sees it. *Biddle v. Perovich* was the 1926 Supreme Court case in which Justice Holmes delivered the opinion that a pardon is not an act of grace, but a "determination of the ultimate authority that the public welfare will be better served" by a pardon.[2] Moreover, the language of Ford's pardon proclamation is strongly reminiscent of Alexander Hamilton's language in *The Federalist*: A pardon "may restore the tranquility of the commonwealth . . . which, if suffered to pass unimproved, it may never be possible afterwards to recall."[3]

This suggests that President Ford took care not to stray from the traditional theory of pardon espoused by the Framers of the Constitution and enunciated by the Court. Ford's pardon proclamation reads, in part:

> Pursuant to resolutions of the House of Representatives, its Committee on the Judiciary conducted an inquiry . . . extending over more than eight months. The hearings of the committee and its deliberations, which received wide national publicity over television, radio, and in printed media, resulted in votes adverse to Richard Nixon on recommended articles of impeachment.
>
> As a result of certain acts or omissions occurring before his resignation from the Office of President, Richard Nixon has become liable to possible indictment and trial for offenses against the United States. . . . Should an indictment ensue, the accused shall then be entitled to a fair trial by an impartial jury, as guaranteed to every individual by the Constitution.
>
> It is believed that a trial of Richard Nixon, if it became necessary, could not fairly begin until a year or more has elapsed. In the meantime, the tranquility to which this nation has been restored by the events of recent weeks could be irreparably lost by the prospects of bringing to trial a former President of the United States. The prospects of such a trial will cause prolonged and divisive debate over the propriety of exposing to further punishment and degradation a man who has already

paid the unprecedented penalty of relinquishing the highest elective office in the United States.

Now, therefore, I, Gerald R. Ford, President of the United States, pursuant to the pardon power conferred upon me by Article II, Section 2, of the Constitution, have granted and by these presents do grant a full, free, and absolute pardon unto Richard Nixon for all offenses against the United States which he . . . has committed or may have committed or taken part in. . . . [4]

Probably for political, rather than philosophical, reasons, President Ford made no attempt to justify the pardon as an act of grace or forgiveness. At first reading, it appears that he was appealing to considerations of justice. For example, he spoke of the wide publicity given the inquiry into Nixon's guilt and implied that a fair trial would be difficult to guarantee. But the import of this is not that a trial would be unfair to Nixon, but that the delay necessary for a fair trial would upset the national "tranquility" so recently "restored." In a similar manner, Ford raised the issue of whether Nixon had been punished enough by losing the presidency. But he was not arguing that the severe 'natural punishment' would make further punishment unfair. The question of what constitutes unfair punishment was important to Ford only because it was a difficult question that would cause prolonged and divisive debate.

So the way Ford cast his proclamation puts the pardon clearly within the tradition of *Biddle v. Perovich*. It was Ford's determination that the public welfare would be best served by letting Nixon retire poolside in peace. The fact that Ford's judgment was arguably mistaken, or that the real effect of the pardon was to allow Nixon to write his own version of history, does not change matters; there is ample precedent for poor or mistaken judgment in pardons. President William Taft, for example, once granted a pardon to a dying man whose release from prison led to a remarkable recovery.[5]

The other dramatic pardon of this period was President Jimmy Carter's 1977 declaration that granted amnesty to all Vietnam War draft evaders.[6] Every war in U.S. history has been followed by pardons and amnesties to reunite the nation and to correct injustices brought on by war-flamed fears. The Vietnam War was no exception; the general amnesty was granted to bind the wounds that an unpopular war had inflicted on society and on its young people, so that healing could begin. But the justification for Carter's amnesty is more complicated than that: Recognition that many draft evaders were doing what they sincerely believed (and were arguably correct in believing) was right raised the possibility that the amnesty might be a way of preventing the punishment of 'innocent' people. If so, this amnesty fits with a different, less established, set of justifications based on principles of retributive justice.

The Atrophy of the Pardon Power

For the past thirty years, an increasing percentage of presidential pardons in the United States have been granted only after sentence has been served. Like his predecessors, Ronald Reagan turned most pardoning decisions over to the Office of the Pardon Attorney, a division of the Department of Justice. "To better reflect the [Reagan] administration's philosophy toward crime," Reagan's Pardon Attorney explained,[7] his office became very exacting in its scrutiny of pardon applications.

As a result, of 3,086 pardon and commutation requests received between 1981 and 1987, President Reagan granted 273 pardons and commuted the sentences of thirteen people.[8] This is a 9 percent rate of approval. In contrast, President Ford, the most generous President, granted 47 percent of the pardon applications he received; and President Franklin Roosevelt, the previous low record holder, granted 22 percent.[9] President Reagan's most generous year was 1983, when he granted ninety-one pardons and two commutations. But the number decreased after that, to a low of twenty-three pardons in 1987.[10] The average number of pardons per year during his administration was 41. This is the lowest average for any president in U.S. history—not just the lowest percentage of applicants or prison inmates, but also the lowest average number of pardons for any president.

The Reagan administration's policy was that pardon applications were not even considered until offenders served their full sentence and waited for an additional five to seven years. Then, if FBI interviews with neighbors, bosses, and attorneys general showed that the offender had led a law-abiding, constructive life after leaving prison, the pardon was *considered*. The offender's reputation in the community, the seriousness of the crime, the need for a pardon, and the prior and subsequent arrest record were weighed in the pardon decision. As the figures show, not many people measured up.

One may well ask what good a pardon does so long after punishment. It stands as a symbol of 'forgiveness,' the Pardon Attorney says,[11] by which he means only that it can be used as evidence that, with the advantage of FBI information and extensive study, the President has judged that the petitioner is clean. A pardon like this does not restore any rights, and certainly it does not relieve many of the painful consequences of conviction. But the pardon can be introduced as evidence in other hearings to restore professional or other licenses that are impossible to get with a criminal record. Moreover, a pardon often helps remove the social stigma of conviction, and it can help in other proceedings to restore the offender's civil rights.

Doubt may be raised as to whether this form of executive clemency—stripped down and hollowed out—even conceptually qualifies as a pardon. A pardon is defined as an act that alleviates or removes the punishment for a crime. Granted so long after sentence has been served and the debt has been paid, this sort of pardon does not remove punishment. One may argue that it removes the memory of the offense or expunges the offender's criminal record, thus relieving the indirect consequences of punishment. But this is not so; a presidential pardon does not, in itself, restore any of the civil or professional rights lost as a result of a criminal conviction.[12] As Pardon Attorney David C. Stephenson plainly admitted, postpunishment pardons are "symbolic" in nature, but as symbols, they "can be useful."

"It's like a high school diploma or college diploma," he explained. "It recognizes that they have achieved rehabilitation and are good people again and recognized by their President as such."[13]

Considering the philosophical underpinnings of punishment in this country in this century, the relative insignificance of pardons in the Reagan administration should come as no surprise. Both the rehabilitative ideal and the new retributivism reduced the role of pardons, in effect and in intent.

The rehabilitative ideal reduced the importance of pardons by giving their job to paroles and indeterminate sentencing. Pardon and parole have been mixed up together for a long time. The first laws granting parole authority were probably modeled on the Governor's power to grant conditional pardons.[14] In Michigan, a successful court challenge to the constitutionality of parole and indeterminate sentences was based on the claim that parole violated the separation of powers by giving prison boards the pardoning power that was rightly vested in the Governor.[15] It finally took a constitutional amendment before parole release was legal in Michigan. Then parole did indeed sharply reduce the uses of pardon, taking many of pardon's traditional jobs on itself. So, on the basis of the rehabilitative ideal alone, the atrophy of the pardon power was predictable.

The new retributivism is doing its own kind of damage to pardon. One of the central recommendations of *Doing Justice* was that there be no early release; prisoners should serve their time. The committee did not come out and say so, but it could readily be inferred that pardons were henceforth an unwelcome intrusion on an enlightened process. Again, while there is no evidence that he did so, President Reagan could easily have read *Doing Justice*, as many get-tough-on-crime politicians did, and (disregarding recommendations for shorter sentences and less incarceration) taken the recommendation against early release to be a confirmation of his own reluctance to pardon ordinary offenders. Law-and-order candidates are not ordinarily pardon enthusiasts, and the new retributivism *appears* to legitimate their position.

The Future of Pardons

The reduced role of pardons has not escaped the attention of scholars.[16] Was this, the second half of the twentieth century, the time for pardons silently to fade away—like collar buttons, their usefulness at an end? This was, to be sure, the ignominious end that both Kant and Beccaria had predicted. And several factors indicated that the end was near. First, pardons had sown the seeds of their own demise, since many grounds for pardon had found their way into the criminal law. The insanity and self-defense pleas are two examples of factors that were once justification for pardon but became part of the criminal law. Second, a primary purpose of pardons—to correct miscarriages of justice—is being served by the increased protection that the Supreme Court has afforded criminal defendants. The courts have taken upon themselves the reassessment of convictions in cases where new evidence suggests a wrongful conviction. The better the justice system gets, the narrower the need for pardons.

This seems to leave only one category of pardons—the pardon granted for non–justice-related reasons, Justice Holmes's instruments of the "public welfare." This sort of pardon came under attack from a different direction, the philosophers. Rather than correcting justice, philosophers Claudia Card and Alwynne Smart independently pointed out, pardons frustrate justice. The philosophers' articles were about mercy, but since they defined mercy as "imposing less than the penalty deserved for an offense," their remarks apply equally to pardon. Smart criticized pardons this way:

> When a man exercises mercy, what he does is acknowledge that an offense has been committed, decides that a particular punishment would be appropriate or just, and then decides to exact a punishment of lesser severity than the appropriate or just one.[17]

If a pardon has the effect of reducing or eliminating a given punishment, and if that punishment was just, then, assuming there is only one just punishment for any given crime, a pardon will have an unjust effect. H. R. T. Roberts agreed: "A genuine act of mercy [or pardon] is always unjustified."[18]

If so, then it is arguable that pardons stand in this position today: If a pardon is used to imperfectly expunge a record, it may not be a pardon at all. If a pardon is used to correct injustices, it may no longer be necessary, this task having been undertaken by other institutions. If a pardon is used as an instrument of the public welfare, it may be unjust. According to Leslie Sebba, "In terms of the clemency power, the effect of this should be the ever-narrowing scope of its ambit, as problems of substance in the penal system are solved by alternate means, and the laws, if not perfected, are at least improved."[19]

Nevertheless, pardons may soon become very important again. It should be remembered that a primary historical purpose of pardons was to make sure that people got the punishment that they deserved so that justice was done. Now, in the name of *Doing Justice* and 'commensurate deserts,' U.S. legislatures are taking steps that will make pardons an essential part of the justice system again. The goal of the new retributivism is to make sure that all offenders get the punishment they deserve—but this goal is to be accomplished by legislated tariffs that set fixed penalties for settled categories of crimes. Since there are only so many levels of punishment, and since the levels of culpability are infinite and the human capacity for causing harm is boundless in its variety, there will necessarily be "hard cases"[20] in which the predetermined sentence is wrong. Then pardon will be the only available remedy.

Moreover, in the federal government and many states, new legislation has guaranteed that parole and other 'good behavior' provisions will soon be a thing of the past. Pardon will be the only way to release offenders from punishment before their time has expired.

It is the central argument of this book that the new retributivism and the reforms outlined in *Doing Justice* will require an expanded and crucial role for pardon. The very same retributivist principles that justify punishment in general, and that determine the penalty that should be imposed, will justify pardon in particular cases. As Sebba pointed out, "ironically, the return to [Kantian retributivism] will thus have ensured the survival of an institution it disliked."[21]

Part II of this book works through the premises and draws the conclusions of this argument by outlining a retributivist theory of pardons. It is a timely effort. As this history has shown, pardon and punishment theories are complementary, and both are responsive to the same political and philosophical ideas. They meet along a dynamic, imperfectly matched faultline. Punishment theory is moving forward; this book is intended to move pardon theory in the same direction.

II

A Retributivist Theory of Pardon

8

How a Retributivist Theory
Can Justify Pardon

The idea that pardons are discretionary acts of grace that need no justification has not served America well. Left to their own discretion, recent presidents have diminished pardons—both numerically and morally. Yet even pardons that are clearly unjust are seen to be exempt from criticism, for who can quarrel with kindness? As a result, when a President grants or threatens to grant a pardon widely perceived as corrupt or unfitting, the disgruntled, grumbling public throws up its hands in resignation. As a first step toward removing pardon's immunity to criticism, it needs to be shown that pardons should be just—and that they are not always so. Also, the question needs to be addressed of what makes a pardon just or unjust.

Part II of this book reexamines the practice of pardon in the context of a retributive theory of justice. My goal is to draw the outlines of a retributivist theory of pardon. I argue that pardons need justification. Further, simple good will does not justify pardons, any more than hatred justifies punishment. Punishment is justified because it is deserved, and only when it is deserved. Pardons too should be granted only when they are deserved. I show how the principles of retributive justice help clarify what it means to deserve a pardon and thus how it is possible to justify pardons on retributivist grounds.

A retributivist theory of pardon?—the phrase sounds oxymoronic. The risk is that the entire undertaking will appear backward or ironic from the start. 'Backward' it certainly is. But 'backward' has the advantage of being a new perspective, one that is sorely needed.

My undertaking is not, however, ironic. It may be historically true that the great retributive theory of Immanuel Kant was energized and empowered by moral outrage at the profligate use of pardon. But it is also true that a retributivist theory can justify pardons.

Pardons Need Justification

It is traditional to think of pardons as conceptually different from punishments. As a result, even though punishment is acknowledged to need justification, pardon often is not. Pardons are presumed to be generally beneficial and so to fall under the principle that there can never be too much of a good thing. Also, pardons are usually understood to be free gifts from the sovereign and, like gifts, to need no justification. Third, since the Constitution does not allow any successful challenge to a presidential pardoning decision, the President need not be concerned about whether his pardoning decisions are justifiable. I wish to challenge this traditional view and the arguments on which it rests.

First, the pardons of this century are a reminder of what Kant knew well: pardons are not always good in themselves and do not always have good effects. Pardons have been used in coercive ways, ways that hurt the person being pardoned—to compel testimony, to facilitate deportation, to force a labor leader to promise never again to lead labor, to frustrate an offender's desire to be executed. Even when the pardon benefits the person receiving it, pardons often do other sorts of damage. A President may use pardons to thwart investigations into his own possible wrongdoing. He might grant a pardon in return for a favor. Pardons granted to people who do not deserve them undercut the moral authority of the President. But enough; it is not necessary to show that no pardons are good. If even a few pardons are malevolently granted or malicious in their effects, that is enough to show that pardons do indeed need to be justified.

When some pardons are so clearly wrong, the rest cannot be assumed to be right.

Second, whatever may have been true of the great monarchies, in the American democracy the pardon is not a gift from the sovereign and cannot be exempt, on that ground, from the need for justification. The Framers granted the pardoning power to the executive alone because they believed that, in one set of hands, the pardoning power would be the most effective tool for preventing injustice and achieving the goals of the state.

They did not make the mistake of assuming that because the President is roughly analogous to a monarch, the President's pardoning power must be strictly analogous to a monarch's "acts of grace." Under the absolute monarchies, any criminal offense was an offense against the monarch. The monarch's powers to punish an offense or pardon an offense both were grounded in his 'ownership' of offenses. In contrast, it is clear that federal offenses are not crimes against the President, but against the people, in whose interests the President acts.

Finally, it is simply a mistake, and an obvious one, to reason that since justification for pardons is not required, it is not possible or desirable. That nothing can overturn a pardon decision offers a powerful temptation to abuse, and argues for, rather than against, the need to offer reasons in defense of a pardoning decision.

If one can strip away all the concepts left over from the seventeenth century—all the "acts of grace" and divine forgiveness—and look at pardons operatively, it is immediately evident that punishments and pardons are closely related. Both punishments and pardons determine how long a person will stay in jail, how long a felon will suffer under the stigma of a felony conviction, how much an offender will pay in fines. It is, of course, imperative to carefully scrutinize the justice of sentences, which means that every punishment must be carefully justified. But since pardons also determine the length of sentences, the same care should be taken to be sure that pardons are justified.

Because pardon decisions are in effect decisions about who will be punished and who will not, and how much and how little, and because the distribution of punishment must be justified by reasons related to justice, pardons too (in general and in particular) must be justified by reasons having to do with what is just. The U.S. Supreme

Court was right to abandon the idea that pardon is a gift or act of grace, immune to criticism or justification.

If pardons need justification, on what grounds can they be justified? My suggestion is that they be justified on the same grounds that govern sentencing procedures under the new reforms—the principles of retributive justice. U.S. criminal law is now undergoing a variety of changes prompted by a rediscovery of (or reaffirmation of faith in) the principles of retributive justice. The new forms of retributivism go by many names—'commensurate justice,' and 'fair and certain punishment,' for example—but all are shaped by a central, enduring claim: Punishment is justified because it is deserved. What sorts of pardons would be justified if pardons were assigned by this same standard?

When all is said and done, it is unlikely that only one kind of reason counts in decisions to pardon, which are complex. But that complexity is what makes it important to sort out different justifications and to be clear about each individually. My goal here is to ask, Given a commitment to retributivist justification, when are pardons justified? When are they not?

Retributivist Justification

Retributivism is, above all else, a theory of justification, a theory that legitimates certain kinds of reasons and denies the moral credentials of others. The traditional subject of retributivist justification is punishment, but its principles can apply to pardon also.

Making clear what retributivism is turns out to be a difficult undertaking. The basic ideas of retributivism are a familiar part of everyday life. People who would not even try to pronounce 'retributivism' are comfortable giving voice to retributivist slogans: "give them what they deserve," "like for like," "a blow for a blow," "measure for measure," "a tooth for a tooth," "tit for tat," and "a dose of their own medicine." But for all their familiarity, the ideas are seldom examined closely. "Tit for tat," people say, without wondering what the words mean.[1]

A major source of difficulty is that there are many varieties of retributivism, as Part I illustrated. There are the moral accounting, religion-based theories; Kant's fairness view of retribution; and Hegel's rights-based retributivism. In addition, there have been many variations on the theme composed in the past few decades, as philosophers rediscovered retributivism and redesigned it to fit their theories and fill their needs.

Central Claims of Retributivism

One thing all retributivist theories have in common—and what distinguishes them from nonretributivist theories—is that they ground justifications in an overall view of justice.

The retributivist claim that desert ought to be the only determinant of how offenders ought to be treated presupposes a set of metaphysical claims. A retributivist believes that there are genuine choices present in the world, that people can rationally choose among them, that people are capable of acting on their choices. In this way, people are free, and their freedom distinguishes them from objects in the world—from dogs and imbeciles and trees. Beyond this, the retributivist believes that persons are political equals unless they choose to make themselves unequal in a way that calls the attention of the state—that is, by committing a crime. A political system and a system of punishment should be consistent with the freedom, rationality, and equality of its members. Moreover, it should respect their essential humanity.

Retributivism also presupposes a basic moral precept, a principle of equity: people who are similar ought (*ceteris paribus*) to be treated similarly. And further, it requires that any difference in the way people are treated ought to be based on significant differences among them. This principle expresses a central idea in Kantian morality: in order to be moral, one must be consistent. Inconsistency goes arm in arm with partiality, and to be partial to oneself or to others is, all other things being equal, unjust.

It is doubtful whether this principle can be supported by argument or that it needs to be. I think that philosopher Richard Wasserstrom was essentially correct when he wrote that "the principles that no person should be treated differently from any or all other persons unless there is some general and relevant reason that justifies this difference in treatment is a fundamental principle of morality, if not of rationality itself."[2]

In regard to punishment, the principle of equity amounts to the claim that people ought to get the punishment they deserve, that only a fact about specific offenders or their offenses justifies a given treatment.

Retributivism legitimates the same kinds of reasons when the subject deserves punishment or pardon: it is just to give people what they deserve. Punishing people who deserve punishment—and not punishing people who do not deserve punishment—is a kind of justice. To a retributivist, that is the only justification that both punishment and pardon need and command.

With this, a retributivist theory sets itself apart from other theories. It sets itself apart, for example, from H. L. A. Hart's theory of punishment. Hart thought that there is only one form that a justification for the general practice of punishment can take; namely, a General Justifying Aim, an explanation of the good that will come of punishment. For Hart, only the prospect that future crime will be reduced can justify the hurt caused by punishment. Hart contended, and I agree, that *if* this is the form that a justification for punishment must take, then retributivism does not justify punishment as an institution.

Retributivists respond that, on the contrary, a General Justifying Aim of punishment does not justify. Why should one person pay the price for future benefits that will belong to others? Retributivists claim that the only fact about punishment that justifies it in general is a General Justifying Characteristic: it is just that people get what they deserve. That is the justification for punishment: *punishment is justified because it is deserved.*

Aspects of Desert

The conviction that people should be treated as they deserve is only the formal beginning of a justification for punishment. People who know that they are to honor this conviction in their actions still do not know how to act. The interesting question becomes,[3] What are the facts that determine what a person deserves?

Retributivists split on this question. Through all the different permutations of retributivist theory, there can be seen two different, identifiable traditions or strains.[4] One is the view that I call "legalistic retributivism": people deserve punishment when, by their crimes, they have gained an unfair advantage over law-abiding citizens. The other retributivist strain I call "moralistic retributivism": people deserve punishment when, by their crimes, they have shown themselves to be morally reprehensible.[5] Although the two strains are identifiable, it should not be inferred that any given retributivist holds one to the exclusion of the other. Most retributivist positions can be found along a continuum connecting the two views, which are sketched separately here in order to make their outlines clearer.

Liability and Desert

Legalistic retributivism defines the considerations involved in a judgment of criminal liability. Moralistic retributivism defines the considerations involved in a judgment of moral desert. Together they define a set of four possible 'degrees of desert,' each calling for a different response from the state:

1. A person may be liable to punishment and morally deserving of punishment. A retributivist viewpoint would insist that punishing in such a case is an unconditional moral obligation. A pardon in such a case would be a serious injustice.
2. A person may be neither liable to punishment nor morally deserving of punishment. Pardon in such a case is an absolute obligation, because the retributive prohibition against punishing people who do not deserve punishment is unconditional.
3. A person may be morally deserving of punishment but not liable to punishment. Since liability is a necessary condition for punishment, punishment is illegitimate in this case also; pardon is called for on retributivist grounds.
4. A person may be liable to punishment but not morally deserving of punishment. In such a case, a pardon is possible on retributivist grounds but not legally required. Reasons having to do with moral innocence have a moral 'heft' and can be weighed in a decision about whether or not to punish.

Thus, retributivism requires pardons when there is no liability to punishment, and it permits pardons when there is liability without moral culpability.

Retributivist Justification for Pardon

Retributivism limits the field of reasons having cogency in a decision about punishment and pardon. The only reasons that are made relevant by the retributivist theory are those bearing on a person's desert. There is a place for pardons in retributive justice, but only in a carefully limited number of cases.

Pardons in Favor of Unfortunate Guilt

Explaining the proper role of pardons in *The Federalist*, Alexander Hamilton wrote that pardons should be granted as "exceptions in favor of unfortunate guilt."[6] A similar idea was expressed by a former United States Attorney General, who remarked that pardons may be proper for "technical violations" of the law.[7] However, efforts to define unfortunate or technical guilt have been awkward.

Some explanations of technical guilt have focused on the intentions of those who formulated the law. According to Smithers,

> If [the case] is one that a dispassionate mind of honest intent is constrained to believe would not have been included in the terms of the violated statute when it was enacted had the legislature known the facts as here imbued with later accepted views, then it is exceptional and clemency should be extended.[8]

This leaves open the question of what would have made the legislature exclude the case from the terms of the statute.

The idea that a person can deserve punishment in one sense without really deserving it in another sense suggests another explanation of technical guilt. A person is only *technically* guilty when he is liable to punishment on legalistic retributivist grounds, but not morally deserving of punishment on moralistic retributivist grounds.

The U.S. government had the right idea in a 1958 handbook:

> There will always be cases of technical guilt but, because of extenuating circumstances, comparatively little moral fault. The pardon power, in other words, is necessary in cases where the strict legal rules of guilt and innocence have produced . . . unjust results.[9]

Philosophers, particularly Smart[10] and Card,[11] have been on the track of this idea for some time. In 1968, Smart wrote,

> [I]f a judge conscientiously examined every case before him, and where the law was too crude and inflexible to bridge the gap between legal and moral justice, exercised mercy, we would probably regard him as a very humane and merciful judge.[12]

The job of Part II of this book is to take this idea and develop it into a set of specific conditions under which pardons are justified on retributivist grounds.

Grounds for Pardons

The chapters that follow explain the theories of legalistic and moralistic retributivism and, then, the corresponding distinction between legal liability and moral desert. From these ideas come a set of specifiable circumstances under which pardons are justified on retributivist grounds.

Liability, as defined by legalistic retributivism, is a necessary condition for justified punishment. In the absence of liability, offenders must be pardoned. Thus, the first ground of pardoning is when this barest minimal requirement is not met:

> *Pardons are justified to override false convictions.*

Insane people or people with substandard mental capacity cannot be held liable, a second ground for pardon:

> *Pardons are justified to prevent the punishment of insane and mentally retarded offenders.*

These pardons for 'innocence' are discussed in Chapter 12.

Liability is not an all or nothing thing, as punishment is not. Sentences of offenders with reduced liability may be correspondingly reduced. Thus, a further ground for pardoning is as follows:

> *Pardons are justified to reduce the sentence of offenders who gained little or nothing by their crimes.*

This sort of legalistic-retributivist justification for pardon is explained in Chapter 13.

Even if offenders are liable to punishment, it is open to a retributivist to pardon them nevertheless when they do not deserve punishment in a moral sense. This is the case when the illegal act is morally justified or conscientious. Thus, a further set of reasons for pardon are these:

> *Pardons are justified for morally justified crimes. Pardons are justified for some 'conscientious' crimes.*

These moralistic-retributivist justifications are explained and illustrated in Chapter 14.

Finally, both legalistic and moralistic retributivism make it important to match the suffering of the offender to the seriousness of the crime. Pardons can be justified to correct a sentence that is harsher than can be justified. Thus,

> *Pardons are justified when the offender has already suffered enough.*
>
> *Pardons are justified when the offender stands to suffer too much because of special circumstances.*
>
> *Pardons are justified to relieve any punishment that is too severe.*
>
> *Pardons are justified to relieve the lingering consequences of criminal conviction.*

These grounds for pardon are the subject of Chapter 15.

9

Legalistic Retributivism

The roots of the views that I have chosen to group under the title 'legalistic retributivism' go back beyond Kant at least to St. Thomas Aquinas. In the twentieth century, a tentative version of legalistic retributivitism appeared in H. L. A. Hart's *The Concept of Law*.[1] In its most recent form, legalistic retributivism is a combination of several ideas of Kant, filtered through the rational contractarian views of John Rawls,[2] and given various forms by Herbert Morris,[3] Jeffrie Murphy,[4] John Finnis,[5] and Michael Davis,[6] among others.

The view presented here under the name of 'legalistic retributivism' is to their views as 'generic' beer is to Budweiser; this view lacks the sparkle and perhaps the clarity of the major theories and should not be mistaken for any one in particular, but it has all the major ingredients. As a result, it can give the reader a general idea of what legalistic retributivism is about. But I should emphasize that the legalistic retributivism outlined here is an extreme version, with its edges drawn sharply to distinguish it from moralistic retributivism. Pure legal retributivism may be a view that no one holds exclusively.

The central insight—the main ingredient—of legalistic retributivism is that punishment is justified (when it is) as part of a larger theory of a just society in which obedience to the law is a primary obligation.

What Is Legalistic-Retributive Justice?

The legalistic-retributivist's view of a just political system is based on a quasicontractual model of reciprocity.[7] The system is established (or maintained) by its members to protect their liberties and to secure their separate rights—and thereby to aid them in achieving their chosen goals. For this to be possible, the members must agree to make certain sacrifices. These are of two kinds. First, members voluntarily agree to give up a degree of *freedom* by binding themselves to obey the laws, even when they do not want to or when obedience is personally disadvantageous. They understand that since others will do so also, their freedom will be enhanced. Second, members assume obligations to contribute what is necessary for the functioning of the state—taxes, military service, and so forth.

As Murphy described it, this understanding of society

> presupposes what might be called a "gentlemen's club" picture of the relation between man and society—i.e., men are viewed as being part of a community of shared values and rules. The rules benefit all concerned and, as a kind of debt for the benefits derived, each man owes obedience to the rules. In the absence of such obedience, he deserves punishment in the sense that he owes payment for the benefits.[8]

A society like this has achieved a sort of balance. It is important to be careful about just *what* is in balance. Is it a balance of tangible benefits that makes a society just? Or is it equal shares of liberty and equal obligations to obey the law that make a society just? Legalistic retributivists do not agree among themselves.

Some legalistic retributivists sometimes argue that a fair distribution of tangible benefits and burdens in a society is a necessary condition that must be achieved before punishment is justified.[9] But this seems to me to be a constraint that is not required by the legalistic retributivist's own premises. To make contrary use of Murphy's image, the members of gentlemen's clubs do, after all, have a variety of incomes and assets, fairly and unfairly gained. So some uneven, or even undeserved, distribution of goods is consistent with justified punishment. If, however, unfairness is so pervasive that 'club membership'—participation in the state—is so seriously and unpredictably disadvantageous that no rational person would seek it under those rules, then the justification for punishing breaks down.

What *is* a necessary condition for the justification for punishment,

I would argue, is that the members of society have roughly equal shares of freedom, expressed as equal obligations to obey the same set of laws. This is because the 'balance' to be preserved is not so much a balance of material advantage—position, power, wealth—as a balance of restraint, that is, a self-imposed restriction on freedom. Society is in balance when all the members are bound by the same set of laws restricting their freedom.

In such a society, crime gives the offender an unfair advantage over law-abiding citizens. The advantage is of two sorts. There is the 'loot,' if such exists, or the material gain to the offenders who may now have something that does not belong to them. And something more important is also gained: an excess of freedom, or, in Finnis's words, "the advantage of indulging a (wrongful) self-preference, of permitting himself an excessive freedom in choosing—this advantage (of exercising a wider freedom and of acting according to one's tastes . . .) being something that his law-abiding fellow-citizens have denied themselves insofar as they have chosen to conform their will . . . to the law even when they would 'prefer' not to."[10]

What Justifies Punishment?

According to this generic form of legalistic retributivism, punishment is justified when it restores fairness by taking away the advantages unfairly gained. It follows from the concept of crime (although legalistic retributivists are not themselves consistent on this point) that since two sorts of unfair advantage are taken by crime, two sorts of punishment are justified. One is reparation; the offender must be forced to give back what was unfairly taken. A Canadian philosopher, Jan Narveson, correctly questioned whether this is punishment, strictly speaking, or merely imposing damages, which could be done as well without the criminal law.[11, 12]

The other sort of punishment is truly punitive. The most important unfairly gained advantage is the wider freedom that comes to those who do not restrain their criminal impulses. Reparations alone cannot remove this advantage. Only punishment can restore the balance of restraint by removing the full measure of freedom the criminal has unfairly gained.

To remove the unfairly won freedom, the offender's will must be restricted through punishment. Finnis wrote that "the criminal has the disadvantage of having his wayward will restricted in its freedom by being subjected to the representative 'will of society' through the process of punishment."[13] Thus, punishment is not essentially inflicting pain; punishment is anything that is forced on offenders by the state, as long as it is not what the offenders would have chosen themselves. Imprisonment is a particularly appropriate punishment— although other punishments could conceivably serve as well—because in prison, the offenders can make almost no choices for themselves.

This is how legalistic retributivism provides one gloss on the retributivist slogan, "Punishment is justified because it is deserved."

When Is Punishment Unjustified?

When are punishments *not* justified by legalistic retributivism? The short answer is, when there has been no unfair advantage gained by violating the law.

First and foremost, no one may be punished who has not broken the law; this is an important guarantee of immunity. Furthermore, legalistic retributivism cannot justify punishing someone who has broken the law but has not profited in any way from the criminal act.

With two possible ways to bring justified punishment down on their heads (by gaining tangible or intangible undeserved advantages), even the most inept criminals can usually succeed on at least one. So illegal acts are rarely excusable. It is true that occasionally offenders commit crimes that do not give any *tangible* advantage: robbing an empty till, picking an empty pocket,[14] performing an illegal abortion on a woman who was not pregnant,[15] shooting with a gun that misfires,[16] and so forth. But the failure to gain a tangible advantage does not, in itself, defeat the legalistic-retributivist justification for punishment. For that to happen, the criminal attempt must not widen freedom, must not bestow the important *intangible* advantage that people gain who refuse to conform their acts to the law—the enjoyment of a wider liberty.

It may be questioned whether it is possible for an offender to commit a crime and take no liberties. Thus, it may also be questioned whether legalistic retributivism can provide a theory of excuses. I think it can, although the lines must be carefully drawn. There are several general types of cases in which it can be argued that a crime takes no advantage:

1. People who act under duress provide the clearest cases of breaking the law without any matching increase in freedom. A woman who commits the crime of prostitution because her boyfriend has threatened to kill her or to deprive her of the drugs she depends on does not have "the advantage of indulging a (wrongful) self-preference."[17] Similarly, but less clearly, a person who is 'driven to crime' by hunger, say, or by drug dependency may not end up with an unfair net gain in freedom.

2. Lawbreakers who are for some reason incapable of choosing to commit a crime do not thereby unfairly widen their freedom. This is true of those who are not able to understand or appreciate what a crime is, through incapacity or insanity. If a person is so feeble-minded, for example, that he thinks that standing guard at an armed robbery is a 'swell game,' he has not "permitted himself an excessive freedom in choosing."[18] Crimes committed under insane delusions or compulsions do not "indulge a wrongful self-preference."[19] There is no doubt that such offenders can gain a material advantage, but that advantage can be removed through the workings of civil law.

Thus, although their edges are not clearly defined, categories of excuses based on duress and lack of capacity or responsibility can be explained by the concept of unfair advantage.

In addition to providing the rationale for a set of excuses, legalistic retributivism can make sense of several other aspects of the criminal law. It explains why nonresponsible offenders may be held liable under tort law, while they are not punishable under criminal law; even an unwilled offense may earn an unfair tangible advantage. It also explains why some factors and not others are mitigating or aggravating circumstances. For example, if a responsible offender has, in the end, gained nothing from his crime, either because he voluntarily made reparations or because a natural punishment resulting from the crime removed his advantage, punishment can be mitigated. Legalistic retributivism also explains why motive may serve as an aggravating or mitigating factor,[20] since the thrill-seeking thief arguably indulges his will more than the hungry thief.

It must not be allowed to escape attention that legalistic retributivism does all this—justifies the institution of punishment on grounds of desert and justifies excuses and mitigating and aggravating factors—without ever making use of the concept of *blame*. No doubt committing a crime is usually unjust, a moral failing. No doubt many offenses and most offenders act immorally. Legalistic retributivists would probably agree. But blameworthiness is not what distinguishes those who should be punished from those who should not. To justify punishment, the legalistic retributivist does not have to show that people who commit crimes are *wicked*, but only that people who commit crimes have gained an undeserved advantage. Whether or not that itself is wicked is a question that need not arise.

Thus, legalistic retributivism is about what offenders *do* and *gain* and only indirectly about what offenders *are* and *take*.

The absence of moral condemnation is no accident. Many legalistic retributivists thought that the old, frowning retributivism had an unpleasant vengefulness about it, the taint of self-righteousness and vindictiveness.[21] The legalistic retributivists cleaned up the concept by stripping away the perceived eagerness to pass moral judgment on the wicked and to inflict pain. That legalistic retributivism succeeds in making blame irrelevant, without losing the justification for punishment *and* the justification for excuses and pardons, is a remarkable accomplishment.

What Is the Legalistic Retributivist Measure of Punishment?

If punishment is justified because it removes the unfair advantage that offenders gain over law-abiding citizens, then the only justifiable punishment is one that does in fact remove that advantage—and no more. Punishment must be commensurate with the unfair advantage; any more is undeserved and therefore unjustified.

As a consequence, several other kinds of factors are irrelevant to sentencing decisions. All forward-looking considerations are beside the point; it does not matter at all what effect the punishment will have on the offenders or the public. The character of the offenders, their lifestyles, their police records, and their good behavior in jail are also irrelevant.

The measure of punishment, in the generic view of legalistic retributivism that I am sketching, has the effect of guarding against three kinds of injustice. First, it guards against the injustice of punishing offenses of the same gravity with different degrees of severity. It prohibits executing one murderer, for example, and giving another a suspended sentence for exactly the same crime. Second, it guards against the injustice of imposing the same punishment for crimes of different gravity, as in requiring both a rapist and a pickpocket, for example, to serve two years in prison. Third, it guards against setting an entire tariff of penalties that are all too lenient or too severe; as would be the case with, for example, a comparatively just tariff that assigns penalties ranging from one cent to a dollar.

The difficulties of deciding exactly what penalty will remove what advantage are enormous. Tangible advantages may sometimes, but not always, be measured in dollars or labor. But how much time in prison wipes out the freedom gained by breaking the law against murder?

Davis should be credited with the only recent serious effort to solve this problem.[22] He suggested that information about the value of the advantages gained by crime could be learned from a sort of auction, in which people would bid for one-time crime licenses (what he calls "pardons in advance"). People would bid against each other, and they would bid against "public-spirited" associations that pool their resources to keep the prices artificially high. The number of licenses for sale would be determined the way Fisheries and Wildlife Department officials determine the number of hunting licenses: by asking how much hunting (lawbreaking) can be done without destroying the prey populations (the society itself).

Davis explained the next step as follows:

> If such an auction included all crimes prohibited by a particular legal system, the auction would yield a complete ranking of crimes according to the unfair advantage taken. That ranking could be set beside a similar ranking of (humane) penalties to determine how much each crime should be punished. The extremes would be settled automatically. The crime ranked highest in unfair advantage would be assigned the highest penalty; the crime ranked lowest in unfair advantage would be assigned the lowest penalty.[23]

Davis' suggestion is difficult and complex. If such an auction were to take place, the prices would probably be affected not only by the fears of public-spirited people (who would buy a license in order to *prevent* crime) and by the selfishness of others (who would buy a license in order to *commit* crimes), but also by the supply of crime licenses in relation to the demand. How much is it worth to people to be safe from crime? How much more is it worth to people to be able to commit crimes with impunity? How much crime can a society tolerate? The auction model makes all these factors count, but does not explain how they relate to one another. So, while it may be a partial theoretical solution to the problem of how to match the seriousness of crimes to the severity of punishment, more clarification is needed. One thing is certain: If a measure of the difficulty of a problem is the lengths philosophers have to go to propose a solution, then this is a difficult problem indeed.

What Are the Limitations of Legalistic Retributivism?

What I have sketched under the title 'legalistic retributivism' is a theory that both justifies and measures punishment. The institution of punishment can be justified by a contractual model in which people agree to obey laws that protect their freedom, in return for the same assurance from others. Since breaking a law grants the lawbreaker an unfair advantage, punishment is justified because it removes the advantage unfairly gained. It is therefore justifiable to impose as much punishment as is necessary to remove the liberties unfairly taken. Thus, no one may be punished who has not broken the law and thereby put himself in an advantaged position.

For all that it accomplishes, the view of legalistic retributivism sketched here has several limitations, all stemming from its exclusive focus on the advantage gained.

First, it should be noted that advantage gained and disadvantage suffered are not necessarily correlative. Since a victim can be seriously harmed by an act that does the offender little good, a punishment could be sufficient to remove the advantage without removing the victim's disadvantage. Moreover, two victims may be harmed to a different degree by the same act. Legalistic retributivism has difficulty accounting for mitigating and aggravating factors based on the relative harm done to the victim or based on the special vulnerability of the victim. It cannot distinguish, for example, the reprobate who steals the last ten dollars from a starving child from the city thug who steals ten dollars from a stockbroker on her way home from work. As long as the advantage is the same, the punishment must be the same.

Finally, this kind of legalistic retributivism focuses exclusively on only one aspect of crime—and perhaps not the most important. A murder does many kinds of wrong. It grants an unfair advantage to the murderer, as legal retributivists correctly point out, who does not limit his own acts to those permitted by law even though he has benefited because others have done so. The murderer has taken liberties and for this he should be punished; so says the legalistic retributivist, and there is no reason to disagree.

But that is the least of it. The murderer has also taken a life. He has done irreparable harm, causing a terrible loss to the victim and many others. In doing so, he has done serious moral wrong. For *this* he should be punished as well. Yet the outrageousness of the crime is disregarded by legalistic retributivism.

This objection amounts to no more than the claim that legalistic retributivism lacks exactly what it was at pains to remove—basing desert on the moral blameworthiness of the offender and the harm done by the offense.

What legalistic retributivism *does* provide, however, is important: the idea that the institution of punishment can be justified by a quasi-contractual model in which people agree to obey laws that protect their freedom and to submit to punishment if they do not; the idea that no one may be punished who has not both broken the law and gained an undeserved advantage; and the idea that the amount of punishment deserved is measured by reference to the crime.

10

Moralistic Retributivism

Moralistic retributivism is the patriarch in the retributivist family tree. The oldest form of retributivism and the parent of most new retributivist theories, it is also the most scorned and neglected. This is the sort of theory people usually have in mind when they criticize retributivism for having an excess of enthusiasm for other peoples' pain, for being an old-fashioned, busybody, vindictive kind of justice. At the same time, it is the theory of justice that probably most closely corresponds to commonsense views of justice, what F. H. Bradley called the "vulgar" view.[1] As the following explanation shows, moralistic retributivism probably deserves neither the vilification nor the popularity. It is a workaday theory of justice that provides a different explanation of "Punishment is justified because it is deserved."

What Is Moralistic Retributive Justice?

The model of justice I am calling 'moralistic retributivism' sees justice in a proportion between virtue and welfare,[2] sinfulness and suffering. Justice is served when people are happy in proportion to their virtue and unhappy in proportion to their moral turpitude. There is a kind of moral alchemy (the sardonic phrase is H. L. A. Hart's) that transforms two evils into a good—or justice. "It is good when bad things happen to bad people," is a blunt, but accurate, expression of this idea.

The main ideas of moralistic retributivism are familiar to everyone.[3] At the end of Westerns, the villain is killed and everybody cheers. The more dramatic his fall from his horse, the louder the cheers. The audience leaves smiling; *that* was a movie with a happy ending. These people are not sadistic; they do not cheer when it is the hero who is shot off his horse. That would be an appalling ending, so deeply rooted is the sense of retributive justice, and so deeply felt is the anger at the universe when it does not follow retributive principles.

This view of justice has a venerable history, and many of its proponents phrase it with grace. G. W. Leibniz explained that the law of justice

> declares that each [individual] participate in the perfection of the universe and in a happiness of his own in proportion to his own virtue and to the good will he entertains toward the common good.[4]

John Rawls formulated the theory this way (in order immediately to discard it):

> There is a tendency for common sense to suppose that income and wealth, and the good things in life generally, should be distributed according to moral desert. Justice is happiness according to virtue. While it is recognized that this ideal can never be fully carried out, it is the appropriate conception of distributive justice, at least as a *prima facie* principle, and society should try to realize it as circumstances permit.[5]

In this view, desert is primarily concerned with the moral qualifications of the recipient whenever a benefit or evil is distributed.[6]

What assumptions support the moralistic-retributivist vision of justice? The principle of equity—justice consists in the similar treatment of similars and injustice in the similar treatment of those who differ significantly—is one of two hidden assumptions. However, if justice requires only the similar treatment of people who are similar, then a system in which all virtuous people are flogged satisfies the requirements of justice.

What is needed to complete the vision is a second claim: As W. D. Ross put it, "a state of affairs in which the good are happy and the bad unhappy is better than one in which the good are unhappy and the bad happy."[7] Further, "what we perceive to be good is a condition of things in which the total pleasure enjoyed by each person in his life as a whole is proportional to his virtue similarly taken as a whole."[8] This "condition of things" is often called 'cosmic justice.'

If justice requires that similars be treated similarly and that happiness be proportionate to virtue, then justice is served when the sinful suffer.

What Justifies Punishment?

One of the purposes of a state is to do justice, moralistic retributivism would argue, in contrast to the 'certified public account-keeping' of the legalistic retributivist. The state is a support system for cosmic justice; by punishing, it makes sure that justice is done in its corner of the universe. If it is just to punish the wicked,[9] and if people who commit crimes are wicked, then people who commit crimes should be punished.

There is a world of philosophy in that little syllogism, along with extravagant opportunities to go wrong. Leaving aside moral issues, one may first entertain the possibility that there is a logical flaw that prevents the argument from getting off the ground. William Frankena thought so.[10] Frankena reconstructed the moralistic argument in the following manner: justice requires rewarding the virtuous and punishing the vicious. Virtue consists in benevolence and justice; moral turpitude is a failure in these same respects. Thus, the just system is one that rewards the just and benevolent and punishes the unjust and malevolent. According to Frankena, such a criterion is circular, precluding its use in the context of moral choice.[11] As long as justice is a part of virtue, then defining justice as rewarding virtue involves a regress that trivializes the definition.

John Rawls makes much the same point:

> The concept of moral worth does not provide a first principle of distributive justice. This is because it cannot be introduced until after the principles of justice and of natural duty and obligation have been acknowledged. Once these principles are on hand, moral worth can be defined as having a sense of justice. . . . Thus the concept of moral worth is secondary to those of right and justice, and it plays no role in the substantive definition of distributive shares.[12]

At least one philosopher, D. C. Emmons, thinks he can refute this argument on its own grounds, since, as he says, "the criterion for virtue can be made to approach a non-circular standard, as the reflexive element is reduced in importance by the nature of the regress."[13] That seems to me to be unnecessary trouble for an uncertain result, since the circle can be broken by the recognition that what is just in a state is not the same as what is just in an individual. The just state is one that punishes moral turpitude, on a moralistic-retributivist view, but it would not be praiseworthy for individuals to take this job on themselves. So virtue in an individual can be defined independently, and the just state is one that punishes people who are lacking in virtue.

A potentially more troublesome premise for any moralistic retributivist is the statement that those who commit crimes are wicked. The premise is, of course, not always true. Some would argue that it is not even *often* true, basing their arguments on the statistical profile of the typical criminal who is a victim of poverty and social injustice.[14] Moreover, breaking the law can be an act of great courage and moral strength.

But moralistic retributivists do not need to deny this. They do not have to be "moralistic" in a pejorative sense. In fact, the moralistic retributivist, more than others, will be most careful not to punish people "just because" they broke the law. People do not deserve to be punished unless their violation of the law was the result of an evil character. The connection between lawbreaking and wickedness is one not of identity, but of evidence: breaking a law is often evidence of moral turpitude, but it is not necessarily moral turpitude itself.

So, the judgment that a given offender deserves punishment is deduced from the judgment that he is a bad person. The judgment that the offender is a bad person is only inductively inferred from his having committed a crime, the hidden premises being that lawbreaking is usually—but not always—immoral and that people who do immoral acts are bad people. Here is an exploded view of the argument:

1. Law breaking is (usually) immoral.
2. People who do immoral acts are (usually) bad people.
3. Therefore, lawbreakers are (usually) bad people.

4. Bad people deserve punishment.
5. Therefore, lawbreakers (usually) deserve punishment.
6. It is just to give people what they deserve.
7. Therefore, it is (usually) just to punish law breakers.

Statement 6 expresses the theory of justification that is common to all forms of retributivism: It is just to give people what they deserve. Statement 4 expresses the central moral principle of moralistic retributivism: Bad people deserve punishment. Assuming that the two retributivist principles are true, the argument makes clear that it will not justify punishment in either of two kinds of cases: when the lawbreaking is not immoral and when there is some kind of disconnection between act and actor, such that the person who does an immoral act cannot be blamed as a bad person.

When Is Punishment Not Justified?

Moralistic retributivism is able to explain under what circumstances it is proper *not* to punish a person (whatever form 'not punishing' takes; i.e., immunity from punishment, excuses, pardons). The rationale turns out to be somewhat different from that justified by legalistic retributivism.

While legalistic retributivism clearly forbids punishing people who broke no law, it is not clear that moralistic retributivism will do so also. If it is just that people suffer when they are wicked, if a person can be wicked without breaking the law, and if it is the government's business to bring about this sort of justice, then nothing in principle prevents a government from punishing moral failings even when they are not also legal failings. Why should greed go unpunished, while theft does not? Why should lust go unpunished, while bigamy does not? If it is not careful, moralistic retributivism justifies too much.

In order to save themselves from intolerable moral account-keeping, it is necessary for moralistic retributivists to postulate that there are a variety of specialized punishing systems simultaneously working toward cosmic justice. Natural justice can be counted on to a certain extent; some forms of debauchery carry their own punishment. God can presumably be counted on to take care of punishing moral wickedness that is not illegal. Social pressures and educational institutions do their jobs of moral education. And governments have been assigned a *limited* duty: to punish moral infractions that are prohibited by law. At issue in the justification of legal punishment is not what people deserve overall, but what they deserve *from the state*. If people act in immoral ways, this may be an affront to the moral order of the universe; however, unless their immoral acts are also illegal, they are not the concern of the state.

This gives no guidance in defining the proper limits to government interference in private life. What immoral acts should be illegal? Once this question is answered, illegal acts are the only sort of acts that are punishable by the state.

This division of labor is a slender argument for something as important as the protection of legally innocent people from punishment. But the fact is that moralistic retributivism blurs the distinction between moral and legal innocence.

Moralistic retributivism is far better at explaining why some people who have committed crimes do *not* deserve to be punished, providing a rationale for a set of justifications and excuses.

First, punishment is not justified, even when breaking the law is wrong, if the offender cannot be blamed for the wrongdoing. So, moralistic retributivism supports several kinds of excuses. It excuses nonresponsible offenders, those who cannot be said to 'own' their acts and therefore cannot be called to account for them—insane or feeble-minded offenders. It excuses those who act under duress, including physical duress, extreme economic duress, and duress caused by disease. All these factors 'count' as excuses because they show that a lawbreaker is not wicked. Since wickedness is the only basis for deserving punishment, punishment is not deserved under these conditions.

Second, moralistic retributivism excuses those who break the law accidentally or because of ignorance or mistake (although only if the accident, ignorance, or mistake is itself not blameworthy). The concept of 'accident' separates an actor from an act. What makes an *act* into an accident is that it does not grow out of the actor's character. Something else—luck—intervenes. A parent may say about a child, "Breaking the vase must have been an accident, because the child is not destructive." Accordingly, accidents are not blameworthy—and not punishable—when they happen in spite of the actor's character.

So far, legalistic retributism and moralistic retributivism, so different in their presuppositions, have produced matching lists of excuses. But there are important differences yet to come.

Punishment cannot be justified by moralistic retributivism in cases where breaking the law is the right thing to do. Even taking into account a *prima facie* moral obligation to obey the law, other moral considerations can be overriding. This possibility allows moralistic retributivism to explain certain justifications that legalistic retributivism cannot. For example, moralistic retributivism argues against punishing people for conscientious acts of civil disobedience; Thoreau arguably should not have spent his night in jail for refusing to support with tax dollars what he judged to be an immoral war.

But what if Thoreau, however conscientious, was simply wrong about the war? It seems to me that the moralistic retributivist would have to say that offenders' sincere beliefs that they are doing what is right lessen (but do not negate) culpability, and thereby lessen the punishment deserved—*if* those beliefs were arrived at in morally appropriate ways. The moralistic retributivist must come up with a standard of what sorts of misjudgments a virtuous person can make and still remain virtuous. Difficult tasks these are, but they are made necessary to avoid the injustice of punishing virtuous people.

The differences between legalistic and moralistic retributivism stand out most clearly on questions of mitigating and aggravating circumstances. The moralistic retributivist must pay attention to questions that are less important to the legalistic retributivist: How vulnerable was the victim? Did the lawbreaker take sadistic pleasure in the crime? How harmful were the effects of the crime? How much was the offender feared? Is the offender sorry? Answers to these questions count because they reflect on the culpability and character of the offender.

What Is the Moralistic Retributivist Measure of Punishment?

The severity of the punishment sought to be proportionate to the seriousness of the moral wrongdoing inherent in the offense. This is the *lex talionis* (literally, the "law of retaliation in kind"),[15] the principle that the punishment should fit the crime. The moralistic retributivist measure of punishment comes from two generally retributivist principles: (1) that justice requires that only those deserving of punishment may be punished and (2) that justice requires treating people unequally only according to the degree of their inequality; and from

a principle proper to moralistic retributivism only: (3) moral wickedness makes a person deserve punishment.

The *lex talionis* presents three distinct challenges to philosophers:
to measure the seriousness of the moral wrongdoing in the offense,
to measure the severity of the punishment, and to match them up so
that one equals the other. These challenges are formidable.

Measuring the Seriousness of the Offense

The first difficulty comes in measuring the seriousness of the offense.
Moralistic retributivism provides the measure in broad outline: moral
evil is the measure for punishment. But an unresolved ambiguity
makes the measure very difficult to apply to cases: Should the punishment fit the evilness of the *act* or the evilness of the *actor*? If the
evilness of the act, is it the evilness of the result that was intended
that counts or the actual harmful effects?

An example shows the ambiguity. Consider an offender who is
a first-class reprobate with a long history of meanness and low-level
violence. Suppose that bullying children is a source of special delight
for him. So he surprises a child on Halloween night and demands her
bag of candy. Suppose also that the terrified child runs into the street
and is hit by a car. Of this entire collection of meanness and misfortune, which provides the measure of the seriousness of the offense?

This sort of question has caused major controversies among those
contemporary retributivists who hold moralistic-retributivist premises. The controversy has generated a large literature,[16] which is rep-
resented here by Andrew von Hirsch, occupying the middle ground,
and by two of his critics, arguing from opposite points of view.

The "commensurate deserts" model advocated by von Hirsch in
Doing Justice takes the middle road, looking to both the act and the
actor to learn how serious a crime is. "Analytically," von Hirsch
wrote, "seriousness has two major components: *harm* and *culpability*."[17] He continued:

> The seriousness of an offense depends, in the first instance, on
> how harmful the conduct is: that is, on the degree of injury caused or
> risked. . . . An offense ought not be deemed serious unless the harm is
> grave. . . . The other major component of seriousness is the degree
> of the offender's culpability: that is, the degree to which he may justly
> be held to blame for the consequences or risks of his act. Here there
> is one well-established principle: that prohibited behavior causing (or

risking) the same harm varies in seriousness depending on whether it is intentional, reckless, negligent, or punishable regardless of the actor's intent.[18]

Repeated crimes are more serious than first offenses, von Hirsch argued, because repeat offenders have persisted in their lawlessness even after they have been warned as forcibly as society knows how. A first crime might be an aberration, but the second crime shows behavior deeply rooted in the offender's character. Thus, von Hirsch believed it important to look to both the act and the actor for a measure of the severity of the offense.

Writing in response to (and in essential agreement with) von Hirsch, Richard Singer advised against weighing the offender's character in a sentencing decision and suggested an alternative scheme. That a determination of 'culpability' is likely to go beyond simple *mens rea* ("guilty mind") requirements, Singer readily admitted: "Why should not other characteristics of the defendant, such as his background, the stresses under which he acted, and the purpose (motive) for which the act was done all contribute to a determination of his blameworthiness? After all, we do, in a normal moral discourse, consider such factors."[19]

But there are at least two excellent reasons for avoiding a judgment of character, Singer said. First, it is simply too hard:

> It may be true, finally, that only God can impose an accurate moral judgment on any person and that to ask judges . . . to make that final judgment . . . is simply arrogant or worse. Is a person who kicks cats more blameworthy for the murder that he has committed than a person who is nice to children? . . . Some questions of morality, after all, are too fine to leave to the state.[20]

The second reason is based on the unfairness of blaming offenders for factors that are—like character—beyond their control:

> Defendants cannot change their past behavior or character and are therefore made liable (or more liable) by events now beyond their control. . . . An analogy to grading of students' exams is apt . . . To give a student an A or F on a C paper because the paper wasn't characteristic would clearly be unfair and unequal.[21]

What Singer suggested instead is that the old common law definitions of crimes be discarded and replaced with a measure of punishment based on "the interests involved or jeopardized *as the crime was carried out.*"[22] Instead of "first-degree armed robbery," for example, an offender would be charged with "crimes in which death or serious bodily harm was purposely inflicted or attempted." By defining crimes in terms of primary interest affected, attention is focused, Singer explained, on "what we are really doing." What we are really doing is, apparently, trying hard to assign the harshest punishments to crimes in which the most important interests were purposefully invaded. The kind of person the offender is turns out to be an issue that can be avoided.

Also writing in response to *Doing Justice*, Martin R. Gardner argued that von Hirsch did not place enough emphasis on the character of the offender:

> Any theory which takes the notions of desert and moral blameworthiness seriously would allow for the careful examination of the whole character of the offender in assessing the issue of his culpability. Consideration must be given to each offender's motives, powers, and temptations if anything like an accurate determination of culpability is to be made.... [I]f punishment really is to be proportionate to the degree of culpability of the offender, the decision of the right amount of punishment to inflict in a given case would entail an assessment of the character of the offender as manifested throughout his whole life and not simply at one weak moment of criminal conduct.[23]

Gardner's suggestion is that the jury should be allowed to look into the whole life history of the offender in order to decide whether, given the offender's character, it is just to punish him for his crime. The jury can then mitigate or even forbid punishment, if it decides the defendant is not blameworthy. One can imagine what Singer's incredulous response to this suggestion might be: A job that can be properly done only by God is now to be done by a jury?

It is by now evident that, in contrast to legalistic retributivism, moralistic retributivism must take into account two sometimes conflicting principles when measuring the seriousness of the crime. The first is this: people can only be punished for what they have *done*. Offenses need to be carefully defined by laws, and people who break the same law should get the same punishment. The second is this: people can only be punished for doing something morally wrong. Legal definitions of crimes, being only a rough guide to blameworthiness, must be supplemented by other factors, and punishments should be carefully individualized.

Measuring the Severity of the Punishment

Compared with measuring the seriousness of a crime, measuring the severity of punishment is a relatively straightforward endeavor. Most contemporary retributivists are generally willing to bow to public opinion on this matter. Davis, for example, suggested the following procedure:

1. Prepare a list of penalties consisting of those evils (a) which no rational person would risk except for some substantial benefit and (b) which may be inflicted through the procedures of the criminal law.
2. Strike from the list all inhumane penalties.
3. [Rank the remaining penalties.] The least penalty is the one any rational person would risk if he had to choose between risking it and risking any other penalty of that type.[24]

This schema is an appropriate starting point, but two further points need to be considered.

First, the severity of a penalty very much depends on the particular vulnerability of the person being punished. Spending time in prison is a harsher punishment for a person who is claustrophobic; working in a license-plate factory is a harsher punishment for a person with arthritis; and so forth. Singer pointed out (correctly, I believe) that a two-month sentence for a sick old man is a life sentence.[25] For the healthy young men of Watergate fame, a two-month sentence was a chance to write a best-seller. Given the differences among people, truly equal punishments would seldom be the same punishment.

Von Hirsch, however, was very suspicious of tailoring punishments to particular offenders.[26] He pointed an accusing finger at judges who use Singer's observations about the different effects punishments have on different people to give mild sentences to white-collar criminals on the assumption that their refined sensibilities will be deeply damaged by even the mildest deprivations. Von Hirsch did not admit the argument that urban offenders accustomed to living with violence and deprivation are hardened criminals, so harsher penalties are required to hurt them since they are accustomed to being victimized. Thus, von Hirsch insisted, it is unjust to fine tune punishments to fit the criminal, even in the name of justice.

But where does that leave proportionate punishment?

A second problem was raised by Gertrude Ezorsky.[27] If punishments are to match the moral evil of the crime, then (disregarding the difficulty of fixing penalties) there is a just amount of suffering that is called for by each crime. But suppose that a given criminal has already suffered dreadfully and undeservedly: maybe (in the clearest case) she has served a jail sentence for a crime she did not commit, or perhaps while she was in jail, her house burned down with her family inside. In a case such as this, the added punishment imposed by the state would be in excess of the pain deserved. Ezorsky called this the "whole life view" of criminal desert, indicating that any measure of deserved suffering has to factor in undeserved suffering already undergone.

W. A. Parent called this approach "patently ridiculous,"[28] and all but the most moralistic retributivists will agree, since the disconnected suffering has nothing to do with the crime. But, it seems to me that the doubt is much harder to dismiss when the suffering undergone is a direct consequence of the crime itself. Suppose a reckless driver backs over her own child, or an attempted murder fails when the bomb explodes in the murderer's hands. There are times when "he has already suffered enough" is true; the criminal act has rebounded to hurt the offender so deeply that justice is served without the addition of state-imposed pain.

So devising a lesser-to-greater scale of penalties is not as easy as it looks. There is tension between the retributivist's need for determinate sentences to eliminate inequities and the retributivist's insistence that the punishment must fit the crime.

How to Match the Punishment to the Crime

The oldest kinds of moralistic retributivism called for the equivalence in kind advocated in the Bible:

> And he that killeth a man shall surely be put to death. And he that killeth a beast shall make it good; beast for beast. And if a man cause a blemish in his neighbor, as he has done, so shall it be done to him; breach for breach, eye for eye, tooth for tooth; as he has caused a blemish in a man, so shall it be done to him *again*.[29]

But Kant made it clear that this was too literal:

The only case in which the offender cannot complain that he is being treated unjustly is if his crime recoils upon himself and he suffers what he has inflicted on another, if not in a literal sense, at any rate according to the spirit of the law.[30]

In recent years, moralistic retributivists have backed even further away from equality, finally settling for what they call 'proportionate equality.' If a relation can be established between one punishment and another in terms of severity, and between one crime and another in terms of gravity, then it is possible to determine a proportion among them such that crime A is to crime B as punishment A is to punishment B. Then it is possible to match the most serious crime with the most serious punishment, and on down the two scales.[31]

This procedure will avoid the unfairness of punishing the same crime differently or different crimes similarly, and thus it accomplishes part of the retributivist agenda; it solves the problem of comparative justice. However, it does not anchor the scales to any absolute standard, and it does not take into account the degrees of difference between crimes. Enrico Ferri argued that

> [n]o scientist, no legislator, no judge, has ever been able to indicate any absolute standard, which would enable us to say that equity demands a definite punishment for a definite crime. . . . If we agree that patricide is the gravest crime, . . . which is the greatest penalty proportional to the crime of patricide? Neither science, nor legislature, nor moral consciousness, can offer an absolute standard. Some say: the greatest penalty is death. Others say: No, imprisonment for life. . . . And if imprisonment for a time is to be the highest penalty, how many years shall it last—thirty, or twenty-five, or ten?[32]

Contemporary retributivists have offered a variety of suggestions. Some invoke a principle of parsimony, requiring that the penalty be as mild as possible, given the goals of the sentencing system. Some argue that punishments should be assigned as if judges were rationing a scarce resource, which, of course, they are. Others are content to insist that the punishments be humane.[33] All of these are good suggestions, but they are not especially retributivist, and they do not touch the problem of determining how much suffering an offender deserves.

I reach the conclusion that moralistic retributivism has an essential truth, missing in every other justification for punishment:

namely, that punishment ought to be related to the moral evil of the crime. It is wrong to punish a person who has done nothing immoral and it is also wrong when a legally and morally blameworthy person escapes punishment. But when it comes to translating this insight into a legal system in which people are fairly tried and sentenced, the problems are enormous and perhaps insurmountable.

What Are the Limitations of Moralistic Retributivism?

Whereas the legalistic-retributivist theory emphasized illegality rather than blameworthiness and paid a certain philosophic price, moralistic retributivism emphasizes blameworthiness rather than illegality and pays a different price. The emphasis on the moral law reduces civil law to a minor role. Breaking the civil law is simply the *occasion for* an inquiry into breaches of the moral law.

One consequence is that moralistic retributivism can give only a halfhearted, this-is-how-we-have-decided-to-do-things justification for what should be a roaring insistence on the connection between breaking the civil law and deserving punishment. A second consequence is that an inquiry into the blameworthiness of offenders has no limits; whether people deserve punishment depends on a host of hidden facts about their motives and their pasts—in short, about their characters as revealed by their whole lives.[34] Finally, fixing penalties is a serious problem.

The upshot is that, no matter what aptitude God may have for rewarding the virtuous and punishing the vicious, this turns out to be a very difficult pattern for running a government. The most serious question is, of course, whether God and governments ought to be in the same business anyway.

But moralistic retributivism cannot be abandoned altogether. It is better than legalistic retributivism at capturing the essence of serious and/or violent crime and the appropriate responses to it. The desire for punishment of violent crime is not based on the resentment (one might say envy) of advantage unfairly taken—"It's not fair that he took what I wanted but restrained myself from taking. He should give back the advantage he took." Instead, the desire for punishment is based rather on moral outrage—"What a terrible person he must be to have done such a terrible thing. He should pay."

11

Deserving Punishment
and Deserving Pardon

According to the retributivist theories outlined in the preceding chapters, punishment is justified because people sometimes deserve punishment and because it is just to give people what they deserve. This is a particularly unhelpful formula. Unhelpful, that is, unless there is some way to know when people deserve punishment—to know what kinds of facts about people, their acts, and/or their circumstances have the moral credentials to justify punishment. Unhappily for any attempt to find a systematic understanding of 'deserving punishment,' retributivism gives two answers—the answer of the legalistic retributivist and the answer of the moralistic retributivist. Worse yet, the two retributivist theories disagree in places, so that legalistic retributivists sometimes justify the punishment of people who would go free on a moralistic view, and vice versa.

So what is a person to do, who wants to assign punishments by retributivist principles? Is it necessary to choose between the two retributivist answers, or do they complement one another?

I suggest that the two sorts of retributivism are complementary. Deciding who deserves punishment requires two different sorts of judgments—a judgment of liability *and* a judgment about moral desert. Legalistic retributivism is particularly suited to the first of these judgments, moralistic retributivism to the second. So, rather than competing, each theory has its own job to do in a two-step process of deciding who deserves to be punished. Working together, they can distinguish among those who *must not* be punished, those who *may* be punished, and those who *must* be punished. Thus, they can define corresponding sets of people who must be pardoned, may be pardoned, and must not be pardoned.

What Does 'Desert' Mean?

The first step in making this argument is to be clear about 'desert.'[1] This step is complicated by a systematic ambiguity: Sometimes "deserves" means "is entitled to," and sometimes it means "is worthy of." Moreover, its meaning changes depending on whether it is a punishment or a prize that is deserved.

Deserving punishment (ill-desert) and deserving a prize (good-desert) cannot be exactly the same.[2] Nonetheless, Joel Feinberg's important book on deserving, *Doing and Deserving,*[3] clarified what it means to deserve a *prize*. This, I suggest, is a promising place to begin. Then, by comparison and contrast, it may become easier to understand what it means to deserve a *punishment* and what it means to deserve a *pardon*.

What Does It Mean to Deserve a Prize?

Feinberg pointed out that, as mentioned above, "deserves" can mean "is entitled to" or "is worthy of," depending on the reasons given for ascribing desert. Feinberg noticed that there is a difference between what he called "qualifying conditions" and "conditions for desert." That is, one can distinguish between those conditions that must be satisfied before a person *qualifies* for a prize and those that must be satisfied before a person *deserves* a prize.

Consider an example: Aaron and Beatrice are involved in an axe-throwing contest. The person who comes closest to imbedding an axe in the bull's-eye will receive a cash prize. Aaron, a novice, slips as he throws and his wildly aimed axe lands exactly in the bull's-eye. Beatrice, an experienced forestry student, loses her grip on the axe handle as she throws. She misses the target completely.

Who deserves to win? On the one hand, Aaron deserves to win because his axe hit the bull's-eye dead center; he satisfied the conditions for getting the prize as they were set down in the rule book. On the other hand, because of Beatrice's superior skill and since the outcome of the contest turned on bad luck alone, one might say that Beatrice *really* deserves to win. The person who has to the highest degree the qualities the contest is to test deserves to win the prize money.[4]

Clearly, "deserves" is used here in two different ways. Following Feinberg, the first sort of desert may be called "entitlement." Aaron is entitled to the prize because he got his axe in the bull's-eye without cheating, which is the only qualifying condition specified by the rules. Beatrice is not entitled to the prize. She missed.

Thus, to be entitled, a person must meet "qualifying conditions."[5] Entitlement is dependent on context. It presupposes the existence of a set of rules that specify the conditions to be satisfied and a mode of treatment that follows their satisfaction.[6] If Aaron and Beatrice are throwing axes outside of the context of a contest (say they are just fooling around on the playground), any claim of being entitled to an award has little sense and no force.

The example of the axe-throwing contest illustrates also the concept of legal entitlement. A person who has satisfied the conditions specified by the law for inheriting an estate, for example, is thereby entitled to the estate. That person has been given title (here literally) and can claim it as a right. The authorities are obligated to enforce this right, to satisfy this claim.

In the second use of the word 'deserves,' people are said to deserve (more often, "*really* deserve") a prize if they have to the highest degree the personal qualities usually required in order to satisfy the conditions of winning the contest. Again following Feinberg, the word "desert" may be reserved for this sort of use. Beatrice *really* deserves to win the prize, because she is the most talented and well-trained of the contestants. Aaron did not really *deserve* to win. He just got lucky.

"Real desert" describes a relation between a person who deserves something and the thing deserved.[7] To ascribe this relation between them involves two judgments of value: first, of the worthiness of the deserving person, and, second, that *in virtue of* that worthiness, the deserving person's getting that which is deserved is a good thing; that in a world in which things are as they ought to be, the deserving person would have what was deserved.

An ascription of 'real desert' calls for reasons. If a person is deserving, it must be in virtue of some characteristic of that person.

It is logically absurd for A to deserve X for no reason at all.[8] Also, not any reason will do.

A condition of worthiness that serves as a reason for saying that someone deserves something may be called, after Feinberg, a "desert basis."[9] It is impossible to specify desert bases in advance or in the abstract because they vary according to what is deserved. But there are several things that may be said in general about these conditions of worthiness:

1. Desert bases are not supplied by any institution or any system of rules. Desert is not mediated; it is logically independent of any institution or any system of rules. Desert is a moral category in this sense.[10]

2. The desert bases must be facts about the people being judged. It makes no difference to a judgment of Beatrice's desert, for example, that her axehead is of the finest Swedish steel, except if that fact indirectly reveals Beatrice's unfailing search for quality.

3. The desert bases must be facts about the past or present. They cannot be predictions about the future.[11] Utilitarian reasons may be good reasons for letting a person win, but they are no reason at all for claiming that the person *deserves* to win.

4. Only an approving judgment can serve as a desert basis for winning something desired. Conversely, only a condemning judgment can serve as a desert basis for ill-desert. Need, ignorance, and illness may be reasons for giving a person something desirable, but they do not serve as desert bases. Likewise, virtue, hard work, and intelligence cannot be reasons for ill-desert. (The one exception is that severe undeserved suffering may on occasion be grounds for reducing the severity of a punishment.) It is unfortunate that philosophers have generally ignored this characteristic of deserving, since it distinguishes judgment of desert from other sorts of judgment.

Other than these four general characteristics, desert bases cannot be specified. This is because they depend on what it is that is deserved.

In summary, it may be said that entitlement is about what people do. Desert is about what people are. Entitlement is about rules, standards, or conditions. Desert is about personal worth or qualities, the old-fashioned concept of "moral desert."[12]

What Does It Mean to Deserve Punishment?

Desert involves the concepts of entitlement and moral worth when what is deserved is an object of *desire*. Desert involves similar concepts when what one deserves is something one would prefer to *avoid*. The case in point, of course, is punishment.

Consider again the axe-throwing contest. Suppose that on her next throw, Beatrice's axe flies far beyond the target and slices off the top of the judge's new hat. Even further, suppose the axe was expertly thrown at that unoffending hat, so angry was Beatrice that she was not awarded the prize she thought she deserved. What does it mean, now, to say that Beatrice deserves punishment? It could mean either of two things.

A distinction can be made between two meanings of ill-desert—namely, liability and moral desert. This distinction is roughly analogous to the distinction between entitlement and desert in the language of winning.

LIABILITY

Punishment, like prize-winning, has qualifying conditions. To "qualify" for punishment, one must commit a crime. The qualifying conditions are specified by descriptions of forbidden acts in criminal statutes. For many crimes, it is not enough to have performed a given act; the statute specifies also the circumstances and state of mind. But whatever the description, a person who may be found guilty of breaking a law may be thought of as having met the qualifying conditions for punishment.

Therefore, qualifying for a prize and qualifying for punishment are alike in being dependent on context. One can only qualify for punishment within the context of a legal system that sets down and enforces standards of behavior.

Rather than bestowing a right, meeting the qualifying condition for punishment removes an immunity.[13] It makes the criminal *liable* to punishment.[14]

MORAL DESERT

The judgment, "he deserves to be punished," may rest not on meeting qualifying conditions, but on a censorious judgment about the accused's character or activities. Just as people use 'really deserves' to convey approval of a competitor who may or may not have a right to a prize, a judgment of desert of punishment can also rest on desert bases. That is, "he deserves to be punished" can mean "he or his actions possess the characteristics that the laws condemn."

As with good-desert, to say that a person deserves ill involves two judgments. The first is a condemnatory judgment about a person's character or actions. The second is the assessment that, in virtue of that condemnatory judgment, punishment of that person would be a good thing.

Moreover, the basis for a judgment of ill-desert conforms to the four characteristics of desert bases previously explained.

1. Criminal laws or penal systems are not a necessary condition for ascribing real desert of punishment. As Kant pointed out, even if all governments were dissolved and their citizens scattered to the far corners of the earth, it would still make sense to say that some people deserved punishment. Even if, when Beatrice sliced the judge's hat, they were the only two survivors of a nuclear war that destroyed all governments, 'she deserves to be punished' makes sense in a way that 'she is liable to punishment' does not.

2. People deserve punishment only by virtue of some fact about themselves—their characters, their acts, their motives. Here again, being liable to punishment and really deserving punishment are different. A man's father murders a neighbor, a woman's husband is a thief, an employer's employee adulterates milk—none of these facts can serve as a basis for desert. A legal system can make these people liable to punishment, but it can never make them deserve punishment. Only a fact about the person punished can do that job.

3. Facts that make a person deserve punishment may not be facts about the future; they must refer to what a person is or has been or what he is doing or has done. Imagine a most extreme case: using sophisticated, computer-aided prediction techniques, a law enforcement agency is able to predict that a teenager who seems normal will grow up to be a vicious predator who murders widowers and squanders pension checks on wanton pleasures. No one could judge on that basis alone that the teenager *deserves* punishment. Only facts about the past and present can do that job. Again, this distinguishes liability to punishment from desert of punishment. It would take only a change in the rules to make the teenager *liable* to punishment; however, only a change in the teenager would make her *deserve* punishment.

4. Finally, only a negative judgment can serve as a desert basis for punishment. This characteristic rules out judgments like these: "He deserves to be punished because he is a skillful lock-picker." Or, "She deserves to be punished because she is a useful informer inside the state penitentiary." Again, these may be reasons to punish, but they are never reasons for saying someone deserves to be punished.

To say that a person deserves punishment is to judge that the person is or has been so wicked that his being punished would be a good thing. This last is the judgment of moral fittingness that is at the heart of moralistic retributivism.

What Does It Mean to Deserve a Pardon?

The distinction between liability and moral desert echoes the distinction between legalistic retributivism and moralistic retributivism. Liability and legalistic retributivism both emphasize the act, the rule violation, and pay little attention to blameworthiness. Moral desert and moralistic retributivism both emphasize the offenders, their characters, and pay little attention to rules.

With all these distinctions in place, it is finally possible to see how a theory that justifies punishment can pinpoint the kinds of cases in which pardon is called for. Retributively justified punishment is based on a determination of desert. Determinations of desert require two steps: a judgment of liability and a judgment of moral desert. At either step, the judgment could be made that a person does not deserve punishment and so should be pardoned. Here is how the decision would proceed:

1. Is the offender liable? This is the question demanded and defined by legalistic retributivism. What did the offender do? Did the offender break the law? Did the offender gain an unfair advantage?

If the answer to these questions is "No," then the offender may not be punished. It would be best, of course, if such a person were never brought to trial. And if brought to trial, it would be best if the laws were written carefully enough that he would not be convicted. But if convicted, he should be pardoned.

Here legalistic retributivism shows its true worth as a limiting principle; it forbids punishment of people who do not deserve it, in any sense of the word. Even an evil reprobate may not be punished for his evilness alone, for he has no advantage unfairly taken. The question of liability is a threshold question: if there is no liability, the offender does not even get in the door.

If, however, the answer to these questions is "Yes, the offender is liable to punishment," then he may be punished since he deserves it in one sense. And then a second question is asked.

2. Is the person morally deserving of punishment? This question asks what kind of person the offender is. Does the offender have the moral characteristics that the law condemns? Did the act that made

the offender liable to punishment also reveal the offender as a moral reprobate? These are questions demanded and defined by moralistic retributivism. And although the judgment must proceed without rules, it must take into account the kinds of considerations supplied by moralistic retributivism.

If the answer to these questions also is "Yes," then punishment is obligatory. For here is a person who deserves punishment in every sense of the word. This person's punishment is fair, it is fitting, and (from the viewpoint of fellow citizens) it is satisfying, the way it is satisfying when the best team wins.

But what if the answer is "No, the offender is not morally blameworthy"? Occasionally a case will arise when a person is liable to punishment, but does not really deserve it. In such a case, the offender may be pardoned.

Thus, retributive justice specifies two roles for pardons in a system of punishment: first, pardons are necessary for people who face punishment even though they are not liable to punishment; and second, pardons are permissible for people who face punishment when they are liable to punishment without morally deserving it.

Observations About Retributive Pardoning

Several observations about pardons follow from this analysis:

First, it should be clear by now how a pardon can be an act of justice rather than an act of mercy. Pardons provide a way to ensure that people suffer only those punishments they deserve, in both senses of the word. Thus pardons provide a way of making sure that justice is served.

Second, pardons sometimes make exceptions to the rules. They can take account of factors that are part of no legal system. The laws forbid acts of a general description because these are generally blameworthy. Pardons are needed when the general presumptions are defeated by exceptional circumstances.

Third, pardons will be few in number in a just society; in a perfectly just society, as Kant foresaw, there would be no pardons at all. If pardons grew to an unmanageable number, one would have to be suspicious that the legal codes were seriously out of kilter with the moral code. As long as a legal system is reasonably just, it will usually be wrong to break the law.

Fourth, because a judgment of moral desert depends on facts that are not often accessible to an executive without invading the offenders' privacy, the burden of proving an absence of moral desert lies with offenders. They are the ones responsible for providing evidence to demonstrate that their crimes were not moral failings. One expcets that this will not be easy to do, nor should it be.

Fifth, it follows from this analysis that it is possible for a pardon to be unjust. On a retributivist view, many occasions for pardon (celebrating Thanksgiving, for example) are left with no legitimate reasons to support them.

Finally, it follows that it is possible to outline the broad categories where pardons are justified; where, in other words, there is likely to be a disparity between liability and moral desert.

12

Justified Pardons: Innocence

Complex systems—nuclear plants, jet airliners—have the potential to do great damage to human life and limb if they malfunction, so they have built-in fail-safe systems as safety measures. For example, valves in a nuclear power plant will shut the entire system down automatically if steam pressure builds to dangerous levels. The computers on a jet have the capacity to override the pilots' judgment to prevent them from making fatal errors. Often there are several layers of safeguards, just in case a primary "fail-safe" system fails.

The legal system is analogous; it is also a complex system with destructive potential. The danger to be guarded against is that someone will be punished, perhaps even executed, who does not deserve to be. The legal system has safeguards to minimize the dangers. They are of two types: (1) the procedural rights of defendants, including rights of appeal, which help prevent innocent people from being convicted, and (2) the detailed descriptions of crimes, which carefully define acts and mental requirements and help prevent the punishment of people who cannot be blamed.

Usually these safeguards work well, but not always. Emergencies arise. The backup system is, of course, the pardon. The pardon provides further assurance that only those who deserve to be punished are punished and only so much as they deserve.

Grounds for Pardon

If pardon is understood as 'free grace,' then there is no need (or way) to specify proper grounds for pardon; pardon can then be granted on "any ground which the executive may regard as sufficient."[1] But when pardon is understood as a way of preventing injustice, then it is possible to specify in advance what sorts of reasons are good reasons for clemency. For once, I am able to agree with Bentham, who wrote that the legitimate grounds of pardoning "are all of them capable of being, and all of them ought to be, specified."[2]

In this and the following chapters, I explain the grounds for pardoning that I believe are justified by the combination of legalistic- and moralistic-retributivist principles. In this chapter are what I take to be the pardons most clearly justified by retributivist principles: those granted because the person facing punishment is innocent of any crime—innocent, that is, in either of two ways. Such pardons can be used (1) to overturn the false conviction of an innocent person, or (2) to prevent the punishment of persons whose abilities are so limited that they are 'innocents,' people incapable of responsible wrongdoing.

In the chapters that follow, I explain three other general occasions for pardon. First are those I group under the heading of 'excuses,' pardons called for because the person facing punishment is not *liable* to punishment on legalistic-retributivist grounds. Second are those pardons grouped under the heading of 'justifications,' pardons called for because the person facing punishment is not *morally deserving* of punishment on moralistic-retributivist grounds. Third are those pardons grouped under the heading of 'adjustments to sentences,' pardons called for to adjust the sentence to inflict only the punishment the offender deserves.

Pardons for False Convictions

Retributivists are absolutely agreed on one point: a person who has not committed a crime may not be punished. So the first and clearest ground of pardoning is when the persons facing punishment are innocent, when they did not commit the crimes for which they are being punished.

It should be noted that, for a retributivist, pardoning innocent people is not discretionary. If a court will not act to prevent the punishment of innocents, then the executive holding the pardoning power *must* act. For a retributivist, liability to punishment is a necessary condition for punishment. If that condition is not met, no other set of circumstances can be sufficient to justify punishment or to bar pardon.

EXAMPLES OF FALSE CONVICTIONS

Does it ever happen that innocent people are convicted? A former District Attorney for Worchester County, Massachusetts, claimed not: "Innocent men are never convicted. Don't worry about it, it never happens in the world. It is a physical impossibility."[3] Unhappily, evidence contradicts the District Attorney. Newspapers uncover accounts of wrongful convictions on a regular basis—but usually not until the hapless accused has spent years in jail. At the urging of Felix Frankfurter, Edwin Borchard compiled a collection of 65 cases in which innocent people were convicted in American courts[4]; his book is a hair-raising account of carelessness, skullduggery, and the inexorable workings of a machine that runs on momentum.

Carelessness. Sometimes false convictions are the result of honest error, zealously compounded. This appears to have been the case when a forty-nine-year-old salesman named Nathan Kaplan was convicted in 1938 for selling heroin to an undercover agent. Kaplan protested his innocence, but three different witnesses convinced the jury he was guilty. A year after Kaplan's imprisonment began, an unrelated man named Max Kaplan confessed to the same crime. He was sentenced to eighteen months in prison. For eighteen months, the two Kaplans were incarcerated in the same Michigan prison, both serving time for a crime that only one of them could have committed. When Nathan was released after a six-year prison term, he filed a motion to set aside his conviction. While agreeing that a terrible miscarriage of justice had taken place, the judge did nothing. "In those exceptional cases where rules of law of broad application work an injustice in an individual case, our institutions provide redress through the pardoning power," he said.[5]

How can something like this happen in a procedurally safeguarded system? Sometimes not all the facts are available, and sometimes available facts are not used properly. Elkan Abramowitz and David Paget explained that there are many ways that fact-finding can go awry: "(1) facts presented to the jury and ignored by them; (2) facts known to defense counsel but not presented to the jury; (3)

suppression of evidence by the prosecuting attorney; (4) doubt as to the reliability of the testimony of key witnesses; (5) failure of defense counsel to make use of an available defense; and (6) failure of defense counsel to develop facts which have been presented to the jury."[6] Armed with facts not available or not used during the trial, the pardoning authority can sometimes prevent the continuation of an injustice.

Skullduggery. Sometimes false convictions occur because of racial bigotry, false testimony, greed, or the working out of grudges. Borchard wrote of Bill Wilson, who married Jenny Wade.[7] In 1908, shortly after the birth of their third baby, Jenny and the baby disappeared. Four years later, fishermen saw bones sticking out of the riverbank. The bones appeared to be the bones of an adult and a child. The fishermen assumed they were Indian relics, but one of Bill's enemies—a man named House—spread the ugly story that these were the bones of Jenny and the baby. Brought to trial for murder, Bill was convicted on the strength of House's detailed and damning testimony. The jury believed House, even though experts testified that the bones could not be Jenny's, and other witnesses testified to having seen Jenny alive and well since her 'murder.' Bill was finally pardoned after three and a half years at hard labor, when Jenny was discovered to be living happily in Indiana with her second husband.

A plot to convict an innocent person was apparently behind the murder conviction of Joseph Smith as well. Fourteen of his descendants asked the Nevada Pardons Board to grant a posthumous pardon to Smith, a union radical convicted of a 1907 murder in a tiny gold mining town in Southern Nevada. The accusing witnesses—"Diamondfield Jack" Davis, a convicted murderer and gunfighter, and "Gunplay" Maxwell, an emeritus member of Butch Cassidy's Wild Bunch—were unreliable. Evidence had turned up indicating that Smith's conviction was a plot by gold mining company bosses. The descendants were gratified by the posthumous pardon, which established that their "grandfather was innocent."[8]

Machine-like efficiency in the legal system. "The law is a mighty machine," Edward Johnes wrote in 1893;

> Woe to the unfortunate man who, wholly or in part innocent, becomes entangled in its mighty wheels, unless his innocence is patent or his rescue planned and executed by able counsel. The machine will grind on relentlessly and ruthlessly, and blindfolded justice does not see that the grist is sometimes stained with innocent blood.[9]

Johnes made the grist mill sound more like a sausage factory, but his point is valid. Prosecution has a forward momentum of its own. In some cases it seems that momentum and the rigidity of rules, more than any viciousness or carelessness, result in a wrongful conviction.

Such was apparently the case of Gary Dotson, who, evidence suggests, was the victim of a deceitful woman *and* the unsavory grinding of the justice system. In July 1977, police came upon dazed and bleeding Cathleen Webb standing beside the road. She identified Dotson as the person who had raped her and slashed her stomach with a broken beer bottle. Dotson was convicted and sentenced to twenty-five to fifty years in an Illinois jail. Eight years later, after having found God, Webb recanted, telling authorities that she had made up the story because she was afraid that she was pregnant by her boyfriend. The judge who had believed her in the original trial did not believe her this time and refused to overturn the rape conviction. Getting convicted was easy enough; getting unconvicted[10] turned out to be impossible. Finally, Illinois Governor James R. Thompson pardoned Dotson—on the grounds that he had served enough time for his crime!

These selected examples demonstrate that, for many reasons, people *are* sometimes wrongly convicted, and innocent people *are* punished—despite the procedural safeguards built into the justice system. The pardoning power is a recognition that human judgment is fallible and corruptible.

IMPLIED GUILT AND THE PARDON FOR INNOCENCE

It might be objected that it is wrong to pardon innocent people because a pardon usually implies guilt, and so those who accept a pardon implicitly admit that they have done something to be pardoned *for*.

Granted: the tendency to infer guilt from pardon is so strong that pardons have occasionally had disastrous results for people who were pardoned. The case of Martin Prisament is a particularly disturbing example. Prisament was convicted of bank robbery in 1937 and sentenced to three years in jail. He insisted he did not rob the bank and could not have done so, because he was in New York at the time of the crime. Two years later it became apparent that Prisament had told the truth; the real guilty parties were apprehended and confessed. Prisament was pardoned by President Franklin Roosevelt on the grounds that he was "innocent of the offense for which he is now being held."[11]

But then Prisament was convicted of attempted robbery. His sentence was severe because he was convicted as a second offender, the pardoned offense counting as number 1.[12]

The conclusion that I would draw from this nightmare is not that innocent people should not be pardoned. Instead, I would urge the conclusion that a pardon does not imply that a person is guilty. Since, on the theory of pardon developed here, fully three out of four legitimate grounds for pardon make reference to some degree of innocence, legal or moral, the imputation of guilt cannot be admitted. A pardon is an official act that reduces or eliminates a punishment. It is more likely to imply innocence than guilt.

If a pardon does not establish or imply innocence, why else would anyone seek a posthumous pardon for a relative? One of the results of conviction is 'infamy'—the offender's name is besmirched, her reputation ruined. When a person has been wrongfully convicted, even long after the unjust sentence has been served, sometimes long after the convict is dead, friends of the wrongly punished person may seek a pardon. Why?—to establish her innocence, to clear her good name, to make sure that her name does not "live on in infamy." So a pardon does sometimes imply innocence. And it usually does so when granted on retributivist grounds.

PARDON AND THE IMPERFECT LEGAL SYSTEM

All the great political philosophers who discussed pardon—Kant, Beccaria, even Bentham—foresaw that improvements in the legal system would narrow the role of pardons until, in a perfect state, there would be no need for pardons at all. The pardon for innocence provides an example.

Before 1907, when the appeals process became a part of the judicial system in England, doubt as to the guilt of the defendant was the most common ground of pardon.[13] In this day and age, however, most judicial systems have some means of "collateral attack"[14] to reopen a case even after all the appeals have been heard; the institution called *revision* in France is an example. In the United States, those who believe themselves wrongfully convicted can appeal on legal grounds, and their right to appeal on the basis of a mistake of fact is steadily expanding.[15]

The result of the liberalized right of appeal has been a marked reduction in the number of pardons granted for reasons of innocence. For the period between 1885 and 1931, W. H. Humbert[16] recorded many presidential pardons given for reasons that may be assumed to relate to doubt about guilt: "disclosure of new evidence," "grave doubt as to justice of conviction," "insufficient evidence," "mental infirmity of judge," "dying confession of real murderer," "mistaken identity," "to rectify mistake in prisoner's commitment," and "innocent." Taken together, these cases represent a significant percentage of the total number of pardons granted. A measure of progress is the number of commutations granted over the past thirty years in Arizona,[17] where "substantial doubt as to guilt" is the only permitted reason for commuting a sentence. The number of commutations granted? One.

Although progress has been made in reducing the number of false convictions, there is still a need for pardons in cases of innocence. There seems to be general agreement that the proper place for a reconsideration of the propriety of a conviction is in the courts.[18] But no one should think that it is an easy thing to get a court to reconsider its decision, even when new evidence arises. The right of appellate review is restricted for reasons having to do with the sanctity of the jury's judgment and the limits to court resources. It is very difficult to challenge a jury's judgment on matters of fact, perhaps a procedural hangover from the Enlightenment idea that "verdict by juries sufficed to exclude every possibility of error."[19] And, because of the widespread view that the law would become too uncertain if there were no time at which a case was fully and finally closed, procedural and substantive blocks make it unusual for a case to be reopened once it has been closed. The general sourness about reopening closed cases came out in an opinion from the California Senate's Committee on the Judiciary. Responding to a man seeking reimbursement for expenses incurred while fighting a wrongful conviction, the Committee said,

> In society it too often happens that the innocent are wrongfully accused of crime. This is their misfortune, and Government has no power to relieve them. It is a part of the price each individual may be called on to pay for the protection which the laws give. He should rejoice that the laws have afforded that protection to him when wrongfully accused . . . and protected him from ignominious death.[20]

Until the judicial system has achieved such a level of perfection that it either makes no mistakes or corrects all the mistakes it makes, the pardon will be an indispensible way to prevent the miscarriage of justice.

Pardons for Reduced-Ability Offenders

Pardon on grounds of innocence is absolute: any person who is completely innocent must be pardoned unconditionally. That is the last that will be seen of firm ground, however, as the discussion moves into the muddier territory where judgments of blameworthiness involve questions of degree and acts of clemency are available in all gradations. The solid rock in this quagmire is, as before, the retributivist insistence that desert (moral and legal) is the only measure of punishment; anyone who cannot be blamed cannot be punished, and if blameworthiness is reduced, so must punishment be.

One reason for reduced blameworthiness—and thus one justification for pardon—is that the offenders have such substantially reduced abilities that they cannot be held responsible for wrongdoing. The offenders, it may be said, are not sufficiently human to be held to account, perhaps because they are insane, or moronic, or children.

INSANITY

In theory, the judicial system makes an attempt to identify insane offenders and spit them out of the penal system into some other social institution or, more recently, onto the street. But defining insanity in such a way that it reliably sorts out blameless people is a problem that has so far defied solution. The drafters of the American Law Institute's Model Penal Code despaired of success:

> No problem in the drafting of a penal code presents larger intrinsic difficulty than that of determining when individuals whose conduct would otherwise be criminal ought to be exculpated on the ground that they were suffering from mental disease or defect when they acted as they did.[21]

Well they might despair: the result of their work was that a person may be found sane—for legal purposes—who, for all other intents and purposes, is a lunatic.[22] Until lawmakers succeed in perfecting a system that protects insane people from punishment, the pardoning executive will have work to do.

Examples are readily available to demonstrate that pardons are still needed. Sara Ehrman collected cases of insane people who were executed for murder. One such case is Billy Rupp, who was executed in California in 1958 for killing a fifteen-year-old babysitter. Ehrman wrote,

> He was a mental defective suffering from a schizoid type of mental disorder, but he "knew right from wrong"—according to some physicians, government witnesses. Another government witness disagreed, "He had a damaged brain, couldn't think for more than a few seconds, a minute, and conceivably didn't know right from wrong on August 8th." Defense medical testimony claimed Billy "was legally and medically insane at the time of the crime." It took the jury 22 minutes to decide he was sane. By 1957 at least six of his jurors had signed petitions to spare his life.
>
> Dr. Henry Sjaarjems, head of the electroencephalograph (brain testing) departments of three Los Angeles institutions and of Camarillo State Hospital, had tested Billy's brain when he was 14. The substance of his findings was that Billy's thought processes were "diffused and degenerated" at 14, and could not have improved at 18.[23]

If Billy's execution cannot be defended on moral grounds, then, from a retributivist point of view, it cannot be defended at all. An act of clemency, rather than execution, would have provided a better match between desert and destiny—always the retributivist ideal.

MENTAL RETARDATION

Diminished-capacity offenders are neglected children in the U.S. penal system. Whether because they are more likely to be caught or more likely to commit crimes, people of subnormal intelligence turn up in the judicial system in disproportionately large numbers. There is no judicial consensus as to how they should be treated. Mentally retarded offenders, if they do not understand their obligations to society, cannot be blamed for violating them. When they are caught in the law enforcement net, their diminished ability is a legitimate reason to reduce or eliminate their punishment.

Examples are sad and sometimes funny: the bank robber who covers his face with a pillowcase but forgets to cut holes for his eyes so that when he stumbles into the bank, gun drawn, he has to lift the pillowcase to peek out—or the bank robber who meekly goes to the end of another line when sternly told by the teller that her line is not for cash transactions. It is a cruel joke, added to the cruel joke played on them by nature, to say that people like this owe a debt to society. What have they gained? How can they be blamed?

Many of the deserters who asked for clemency after the Vietnam

War were mentally retarded. Number 1 on the Clemency Board's list of mitigating factors was the "inability to understand obligations or remedies." The Board was reluctant to send AWOL soldiers to jail if their IQ did not reach 80.

Ohio Governor Michael DiSalle told of how he had to decide whether or not to commute the sentence of Lewis Niday, who had been sentenced to die in the electric chair in 1959. Niday's IQ was 53.[24] He was a big, dull-witted man who had the misfortune of falling completely in love with a married woman "of a certain animal attractiveness," as the Governor said. Through a complicated series of events, taking place mostly in bars and bedrooms, Niday ended up killing the woman of animal attractiveness and the new lover she had attracted. In a letter to Governor DiSalle, Niday's grade-school principal laid out the moral problem posed by Niday's death sentence:

> Lewis was . . . a nervous boy of very low moron type. He passed the first grade only in two years. He never passed another but was passed on every two years on size . . . He gave me no great trouble but I could teach him but very little as he was only half there. I don't think he was ever fully sane. He never had a mind more than a seven-year-old, with no judgment. Please do what you can to save his life as he was never accountable for his acts.[25]

The Governor responded by permanently postponing Niday's execution. "It is true that he killed," said the Governor, "but for society to kill him in return would be no more than an act of vengeance."[26] This is good retributivist philosophy: What distinguishes a punishment from an act of vengeance is that the punishment is called for by legal and moral blameworthiness, while vengeance is exacted without regard to desert.

YOUTH

If an adult who has only the capacities of a child deserves clemency, all the more does a child—who has only the capacity of a child—deserve clemency. A recent spate of death penalties pronounced against violent teenagers has raised important questions: What sorts of capacities are presupposed by the inference from lawbreaking to blameworthiness? What capacities may be presumed in a child?

The difficulties of these questions probably account for the unsettled nature of the law regarding teenage offenders and the mixed emotions of fear and protectiveness that arise in response to a small murderer. But public sentiment is strong that children cannot be expected to have the self-control of adults or to fully understand the enormity of their crimes. Until the legal system figures out a way to protect children from adult-sized punishments, this will be an appropriate role for the pardoning executive.

People innocent of any crime and people too 'innocent' to be blamed for the crimes they have committed have this in common: they do not deserve punishment, and so they should not be punished. It is not only philosophers of a retributivist bent who believe that; legislators and judges must believe it also, since the judicial system is generally set up to protect such people from punishment. When the system malfunctions and the protection is imperfect, the use of the pardon is an important safeguard.

13

Justified Pardons: Excuses

The good reasons for pardoning grouped here as 'excuses' are justified by the principles of legalistic retributivism. To see how legalistic retributivism justifies pardons, consider the straight-forward fact description of a crime, provided by this familiar rhyme:

> The Queen of Hearts, she made some tarts,
> All on a summer's day.
> The Knave of Hearts, he stole those tarts,
> And quickly ran away.

How would a legalistic retributivist analyze the situation? What, on legalistic retributivist grounds, makes the Knave punishable? What possible set of changed circumstances would make him pardonable?

The Knave and the Queen—and all the other Hearts—may be taken, for the sake of argument, to be members of a society that is held together by principles of reciprocity. All agree to limit their own freedom by obeying a set of rules. (Of course, Heart society is not *really* like that, a fact that may or may not be beside the point.)

What is then objectionable about the Knave's act is that he has broken one of those rules. Thus, he has taken unfair advantage of the other Hearts. His advantage is both tangible and intangible: he has the tarts, and he has increased the relative measure of his freedom by not conforming his acts to the rules that others obey. He has taken the tarts, *and* he has taken 'liberties.'

The Knave must be punished as severely as is necessary for the balance of advantage to be restored. The punishment will remove both advantages: the tangible advantage, by forcing the Knave to give back the tarts or some equivalent, and the intangible advantage, by forcing him to submit to some treatment against his will.

Now it is possible to imagine that the Knave took the tarts and gained no such unfair tangible advantage. Maybe he tried to take the Queen's tarts, but someone else got there first and there was nothing left to steal. Maybe he took the tarts, but was immediately stricken by a twinge of conscience so acute that he replaced them without a taste. Maybe the tarts were poisoned, and he inadvertently saved the Queen's life at the cost of his own. Maybe the tarts rightfully belonged to him, having been seized by the Queen's guard as an unfair and extortionate tax. Any of these would be good retributivist grounds for reducing his punishment.

On the other hand, one may imagine that the Knave gained a tangible advantage, but no intangible advantage—no unfairly widened freedom. Maybe he acted unintentionally, being so myopic that he mistook the tarts for his own hat, or maybe he absentmindedly picked up the tarts as he swept up the courtyard. Maybe he was forced to steal the tarts by a passing band of ruffians. Again, any of these would be good grounds for reducing his punishment.

One may even imagine that he gained nothing at all, neither tangible nor intangible, picking up the tarts by mistake and replacing them as soon as he discovered his error. This would make his crime pardonable.

The troubles that real humans get into are every bit as varied and unpredictable—even absurd—as the Knave's tangle with the law. The varieties of human misfeasance are so numerous and unpredictable as can scarcely be imagined, let alone encoded in the criminal law. And, in truth, sometimes real humans break the law and gain no advantage.

When that happens, legalistic-retributivist principles would call for a reduced punishment or, in some extreme cases, no punishment at all—cases like these do not *deserve* punishment. But because no law can anticipate the endless variety of exceptions, the institution of pardon is the way that a state can, in Alexander Hamilton's graceful phrase, "make exceptions in favor of unfortunate guilt."[1]

Accordingly, legalistic retributivism can justify clemency or reduced sentences under two sorts of conditions: (1) in cases where there has been no tangible gain, as in unsuccessful attempts, crimes where the advantage has been returned, and crimes that compensate the victim for unfair treatment; and (2) in cases where there has been no intangible gain, as in unintentional crimes and coerced crimes. Moreover, when a crime confers no unfair advantage at all, legalistic retributivism can justify an unconditional pardon.

Crimes Without Tangible Gain

Legalistic retributivism justifies punishment to the extent that it restores the balance of advantage that was disrupted by a criminal offense. If no disruption took place, punishment has no work to do and cannot be justified. Some crimes are absolute failures; they do nothing for the criminal. Others are partial failures. One sort of partial failure occurs when the criminal gains nothing from a crime except for the extra liberty of not conforming his actions to the law. Because less advantage is taken, less punishment is justified. These are the crimes that a legalistic retributivist would agree call for reduced sentences.

Unsuccessful Attempts

There are many ways to fail in an attempt to commit a crime. Pure accident can make the crime a failure, as when a bullet misses the intended victim.[2] An attempt can fail because it is legally impossible to commit that particular crime, as when a husband attempts to 'rape' his wife in a state where that is not a crime.[3] Or the crime may be factually impossible, as when the offender offers a bribe to a person whom she believes to be a juror but who in fact is not,[4] or when the offender, erroneously believing the gun to be loaded, points it at her husband's head and pulls the trigger.[5] Also, the opportunity to commit a particular crime may simply never arise.

These cases have in common that in each the *intention* of the unsuccessful criminal cannot be distinguished from the intention of the successful criminal. This might suggest that the two are morally indistinguishable.

Consider an example suggested by philosophers Gerald Dworkin and David Blumenfeld: Mr. A decides to assassinate the President. He buys a gun, studies the parade route, and rents a room along the route where he hides the gun. His alarm fails to ring on the fateful morning, and Mr. A oversleeps, missing his chance.[6] Contrast Mr. A to Ms. B, who is identical in all respects except that she has a reliable alarm clock and so succeeds in killing the President. Are they not equally blameworthy?

Brand Blanshard argued that they are morally indistinguishable:

> A man who tries unsuccessfully to kill another is on the same ethical level as if he had succeeded, but he is not treated in the same way. In fact, magistrates have been known to postpone decision on a prisoner until it is learned whether his victim is going to get well or not.[7]

Dworkin and Blumenfeld raised important questions about blameworthiness: "Why do we differentiate individuals with respect to assigning punishment when our moral assessment of them is the same? . . . Why don't we punish individuals for bad intentions when the only factor preventing them from translating these intentions into actions is lack of opportunity or accident?"[8]

I do not believe that either Blanshard's or Dworkin and Blumenfeld's answer is satisfactory. Blanshard explained that "the law in such cases reflects the difference in public indignation toward a murder as against a mere attempt at it."[9] But what, it must be asked, explains the different levels of indignation?

Dworkin and Blumenfeld are even less helpful:

> [N]o ethical theory discussed is capable of giving a satisfactory account of why we always punish attempted murder less severely than murder. In particular, no theory gives an adequate explanation of cases in which accident accounts for the failure of an attempted crime.[10]

But they reckoned without legalistic retributivism, which does provide a moral basis for distinguishing failed attempts from successful crimes. If liability to punishment is measured by the degree to which an act actually upsets a fair apportionment of rights and duties, then a failed attempt is not fully punishable. Although it imparts an intangible advantage and is punishable *on that account*, an unsuccessful crime is otherwise unpunishable.

Laws usually prescribe milder punishments for attempts than for completed crimes. The implicit acknowledgement is that when two offenders' intentions are the same—as they are with Mr. A and Ms. B above—their *intentions* are equally blameworthy and so the offenders are equally liable to moral censure. But their *acts* are not equally blameworthy and so the offenders are unequally liable to legal censure. In contrast to moralistic retributivists, legalistic retributivism does not seek to match punishment to moral desert.

A distinction made by Kant underscores the point. He recommended distinguishing between *juridical* duties and *ethical* duties.[11]

> All duties are either duties of justice (*officia juris*), that is, those for which external legislation is possible, or duties of virtue (*officia virtutis s. ethica*), for which such legislation is not possible. The latter cannot be the subject matter of external legislation because they refer to an end that is (or the adoption of which is) at the same time a duty, and no external legislation can effect the adoption of an end (because that is an internal act of the mind), although external actions might be commanded that would lead to this end, without the subject himself making them his end.[12]

All citizens ought to fulfill their duties of virtue. But only duties of justice are enforced by law. Legal systems punish violations of a duty of fairness, that is, "the active violation of the rights of others, . . . the breach of juridical duty."[13] Even though a violation is intended, punishment is not called for if no rights are actually violated. It is likely, however, that punishment will be called for to remove the widened liberty that the unsuccessful criminal has taken.

Repaired Crimes

Historically, atonement, the "reparation or satisfaction made for a wrong or injury,"[14] has been a common reason for granting pardon. Thieves are often extended clemency when they return the stolen property. In 1913, the United States Attorney General recommended a pardon for a fallen bank officer who "sacrificed his entire fortune, before indictment, in order that the creditors of the bank might be paid, and was untiring in his efforts to aid the authorities in straightening out the affairs of the bank."[15] Political science professor James Barnett made reference to cases in which offenders won release by atoning for their crimes in a peculiar way: they married the women they had forced into prostitution and thus presumably removed the damage done.[16]

Except for the last example—which may be optimistically discounted as an anachronism—these acts of clemency can be justified by legalistic retributivism. If offenders have completely paid back the material advantage unfairly gained, then they cannot be required to make further reparation, although they may be punished to remove the liberties taken.

Crimes of Compensation

What is to be done about crimes that create, rather than destroy, a just balance of advantages? Here is a simple example: C steals D's umbrella, and now has two, when before she had one. Loath to get wet, D steal's C's umbrella. Now they each have one. C's crime is punishable. But what about D's? Although the acts are externally the same, D's 'theft' redressed a wrong; it removed an advantage unfairly won. There is strong public sentiment, arguably buttressed by legalistic retributivism, that a 'crime of compensation' like this does not call for punishment.

But how far can this argument be taken? Suppose that C is a coal mine owner and sole employer in a sooty hilltown. Suppose that C employs D at unfairly low wages, skimming the profits of D's labor for himself. If D steals from her employer, does legalistic retributivism require that she be punished? Or does it require that she be pardoned?

Can the argument be taken further? The Knave of Hearts does not really live in a society bound by rules of reciprocity. He lives in a monarchy, in fear of the Queen who (when she is not baking) is cutting off peoples' heads without justification. Does the Knave owe a debt to the Queen, or she to him? If she owes him, how can any punishment be justified?

Murphy has argued[17] that the legalistic-retributivist justification for punishment presupposes an ideal society that is so far removed from actual societies as to make the theory inapplicable and dangerous. Murphy elaborated:

> Punishment as retribution (paying a debt to one's fellow-citizens) makes sense with respect to a community of responsible individuals, of approximate equality, bound together by freely adopted and commonly accepted rules which benefit everyone. This is an ideal community, approximating what Kant would call a kingdom of ends. In such a community punishment would be justly retributive in that it would flow as an accepted consequence of accepted rules which benefitted everyone (including, as citizen, the criminal). But surely existing human societies are not *in fact* like this at all. Many people neither benefit nor participate but rather operate at a built-in economic or racial disadvantage which is in fact, if not in theory, permanent. The majority of criminals who are in fact punished are drawn from those classes, and they utterly fail to correspond to the model which underlies

the retributive theory. Surely we delude ourselves in appealing to the retributive theory to justify their punishment.[18]

Statistics about criminals confirm Murphy's claim that the majority of crimes are committed by economically underprivileged people.[19] A presidential commission on law enforcement confirmed this characterization of the criminal:

> From arrest records, probation reports, and prison statistics a "portrait" of the offender emerges that progressively highlights the disadvantaged character of his life. The offender at the end of the road in prison is likely to be a member of the lowest social and economic groups in the country, poorly educated and perhaps unemployed.... Material failure, then, in a culture firmly oriented toward material success, is the most common denominator of offenders.[20]

These statistics may be disputed on the ground that they measure criminals caught and punished rather than crimes committed,[21] Murphy admitted. Affluent criminals may be missing from these statistics because the better-educated criminal is less likely to be caught, and once caught, the wealthier criminal is less likely to be successfully prosecuted. This inequity, however, confounds even further the claim that punishment is justified because the criminal owes something to society.

In general, the criminal is overwhelmingly disadvantaged—he is poor, he is uneducated, he is often unemployed. Further, the criminal is *unfairly* disadvantaged; his poverty is often the direct result of laws that are themselves unfair. As Murphy explained,

> [The average offender is an] impoverished Black whose whole life has been one of frustrating alienation from the prevailing socio-economic structure—no job, no transportation if he could get a job, substandard education for his children, terrible housing and inadequate health care for his whole family, condescending-tardy-inadequate welfare payments, harassment by the police but no real protection by them against the dangers of his community, and near total exclusion from the political process.[22]

In consequence, the justification for punishing the already disadvantaged criminal because he has "broken rules set up for the equal benefit of all" is an unsound argument and a cruel deceit. So Murphy argued, challenging the legalistic-retributivist justification for punishment.

How much damage does his argument do?

First, I would point out (as Murphy noted) that the argument does not damage the legalistic-retributivist justification for punishment *in theory*. Support comes from Kant, who argued that an idea can be true in theory, even if false in practice.

> Nothing, indeed, can be more injurious, or more unworthy of a philosopher, than the vulgar appeal to so-called adverse experience. Such experience would never have existed at all, if at the proper time those institutions had been established in accordance with ideas, and if ideas had not been displaced by rude conceptions which, just because they have been derived from experience, have nullified all good intentions. ... This perfect state may never, indeed, come into being; none the less this does not affect the rightfulness of the idea, which in order to bring the legal organization of mankind ever nearer its greatest possible perfection, advances this maximum as an archetype. For what the highest degree may be at which mankind may have to come to a stand, and how great a gulf may still have to be left between the idea and its realization, are questions which no one can, or ought to, answer.[23]

Second, I would admit that injustice in the legal system *does* do damage to the legalistic-retributivist justification of punishment *in practice*. But, how much damage?

If one believes that society is rotten *to the core*, then punishment is always an instrument of injustice—and it cannot be defended on moral grounds.

But it is important to clarify just how unjust a society has to be before punishment becomes merely violence. The society's criminal law would have to be worse than no law at all,[24] running on fear, enforced by raw power. This does not describe American criminal law, despite its profusion of faults. If one believes, as I do, that some laws and practices are unjust and others just in this struggling society, then one may agree with critics that there are cases in which punishment is not justified—while nonetheless disagreeing that *all* cases fall into this category.

Even in an unjust society, offenders can commit crimes that unfairly disadvantage others. Even (or especially) in a Nazi prison camp, it was unjust for one condemned person to steal food from another. The difficulty becomes how to distinguish crimes that take unfair advantage in an imperfect society—and thus deserve punishment—from those that do not take unfair advantage—and so should be somehow exempt from punishment.

James Sterba[25] suggested an answer: A person does not deserve punishment on what may be called legalistic-retributivist grounds when his criminal activity meets two conditions; namely, when it

1. is undertaken after other reasonable options for achieving reasonable progress toward a just society have proven ineffective or too costly for those whom they are intended to benefit and . . .
2. involves only minimal violations of the moral rights of others.[26]

On these grounds, most crimes are punishable even in an unjust society. But some crimes do indeed have the effect of restoring what was unjustly taken. In those cases, the legalistic-retributivist justification for punishment fails. These are the kinds of cases in which pardons are justified.

Pardon can thus be an agent of reform, as it has often been in the past, in two ways: Individual pardons can, in the first place, blunt or thwart the enforcement of unjust laws in individual cases. Second, the accumulated weight of pardons can drive reform, focusing attention on the unjust law and bringing pressure to bear against it.

The operation of the draft during the Vietnam War is an apt example. There is no question that the draft worked unjustly. The large majority of draftees were poor, undereducated members of minority groups. Many were "the ones who were already the victims of life's inequities and society's discrimination."[27] People with connections, with money, with education, could often find a way to escape service.

Everyone who was of draft age during that time has stories of how connections or influence aided the evasion of the draft; many of the stories are autobiographical. Here is a typical story: A barber in a medium-sized California community had two sons. "As the boys reached draft age, their father worried about their being called. One of his customers was a colonel in the Army Reserves. The barber asked the colonel if there was anything he could do. One week later, the boys were jumped to the very top of the list for their local unit. They were soon Reserves. They never went on active duty, and they never went to Vietnam."[28]

After the Vietnam War, there was a remarkable lack of enthusiasm for punishing draft evaders and deserters. The arguments for immediate amnesty had a retributivist color: "Who can call for justice in the name of a law which epitomized the perversion of justice?"[29] Pardons and amnesties were forthcoming as, on legalistic-retributivist grounds, they should have been.

If one accepts that society is fundamentally and irredeemably unjust, the philosopher's job is relatively easy: almost no punishments are morally justified, and everything that goes under the legitimating name of 'punishment' is simply the hurtful use of power. This was the reasoning of leaders in Germany before and after World War II, when each successive government emptied German jails on the grounds that the laws under which the criminals were sentenced—the laws of the preceding government—were unjust.[30]

But if one believes that some laws are unjust and others just, a terrible philosophical job lies ahead: that is, to sort out the crimes that take unfair advantage from those that reclaim advantage unfairly taken, tailoring punishments to remove only unfair advantage. All this must be done while keeping in mind that a person who is un-punishable because the advantage taken is advantage owed may be punishable for unfairly taking liberties. Pity the distracted Governor who is called on to make such narrow judgments.

But, while many judgments of that sort will be difficult, there will be clear cases of deserved pardons. When a criminal is clearly "more sinned against than sinning," a Governor could grant a pardon and justify the pardon on these legalistic-retributivist considerations.

There is one final retributivist consideration that buttresses the reluctance to punish crimes that redress rather than create an injustice; namely, the idea that every crime has a particular, correct punishment. Punishing a person for a crime against a system that has already inflicted suffering on him causes him a total of more suffering than he deserves on account of the crime. Jean Valjean, the starving protagonist of *Les Misérables*, serves as an example: from despair at a hopeless condition, he commits a crime that he would not commit under different circumstances. In this case, as Card pointed out, "it can seem here that the offender has already 'paid' in advance for his offense, although not 'paid' willingly . . . ," and there are limits to the suffering we can impose on a person who is already in despair, without being ashamed.[31]

Crimes With No Gain At All

Even though an offender gains nothing from a crime for any of the reasons just described, the offender may still deserve punishment. The bare fact of having committed a crime gives an offender an advantage over other citizens who control themselves, an advantage measured in freedom of action. But even this advantage is not gained when an offender breaks the law but "didn't mean to do it," either because the act was unintentional or because the criminal act was coerced by other people or by circumstances the offender could not control.

In Anglo-American law, there is a great reluctance to punish anyone who did not mean to break the law. The absence of a "guilty mind" is usually an excusing condition, generally some circumstance that makes the agent incapable of choice or of carrying out what he has chosen to do.[32] And "necessity" usually justifies an act that would otherwise be criminal. But the law is not always as careful as the retributivist might wish. When the full measure of punishment is excessive, and the offender has not taken a full measure of unfair advantage, clemency is justified.

Strict Liability Offenses

One area in which retributive justice would call for a *reduced* sentence is the area of strict liability offenses. Crimes of strict liability impose punishment on persons who may never have intended to break the law and may not even have been aware that they were doing so. The paradigmatic case is the hotel porter who carried a guest's suitcase to his hotel room. Unbeknownst to the porter, the suitcase contained a bottle of alcohol. The porter was convicted of the strict liability offense of transporting intoxicating liquor.[33] Had the porter brought his punishment on himself by taking liberties he would not allow to others? Davis's example was Keith Baker, whose cruise control stuck, causing his car to speed down the highway, catching the attention of a highway patrolman. Can Baker, who may be assumed to be blameless, be punished?[34]

When an offender does not know that an act is illegal or that what she is doing is the sort of act that she knows to be illegal, or if the offender does not know what she is doing at all, it is problematic to say that punishment is called for to remove the extra measure of liberty that she exercised—for it is difficult to say what the offender has gained.

Davis believed that it is possible, albeit difficult, to define the advantage gained by strict liability offenders.[35] Being able to race down the highway with a stuck cruise control *with impunity*, for example, is a liberty some people would value. Even though the hapless offender is blameless—not having chosen to take liberties—he is still in possession of the liberties he unintentionally took and so may be punished on that account. However, the punishment should be less, since the advantage is less. Davis pointed out that,

> if crimes of strict liability can be shown to take some advantage, but less than the advantage taken by the corresponding intentional, reckless, or negligent crime, it would follow that such crimes deserve some penalty, albeit a relatively light penalty.[36]

The same impulse to reduce the penalties of people who are blameless moved a Massachusetts Governor, who went even further than legal retributivism recommends. In 1844, he pardoned a man who had been convicted of bigamy—a strict liability offense in most of the fifty United States.[37] The hapless bigamist had remarried after he was convinced (justifiably, but wrongly) that his first wife was dead.[38]

Coerced Crimes

Not very many people are ever forced to commit a crime with a gun in their back, but it does happen. More often, people commit crimes while under the indirect control of other people, as was the case in Guyana, when, under the influence of the Reverend Jim Jones, mothers poisoned their babies, friends killed friends, and all the "survivors" then killed themselves. The 1970s spawned a string of strange cases of (allegedly) coerced crimes: Lieutenant Calley's murder of Vietnamese civilians, Patricia Hearst's participation in an armed robbery, and the murders committed by young women under Charles Manson's control. Terrible crimes, all of them—but if they were really done under someone else's control, then they are not fully punishable. A retributivist could make a good case for clemency on the grounds that the offenders did not gain an advantage over fellow citizens by indulging their criminal preferences while others restrained theirs.

If retributivism calls for clemency in cases of physical coercion, then can punishment in general be justified if people are metaphysically unfree? If people are simply automatons who move as they are pushed by physical or psychological forces, then perhaps the Utilitarians are correct in saying that persons should be either manipulated in the public's best interest or treated in their own best interests.[39]

Punishment of 'automatons' is a gratuitous cruelty rather than a just desert.

The question of human freedom is crucially important to retributive theory. It presupposes that those who choose to commit crimes generally turn down an opportunity equally open to them to obey the law.[40] Exceptions to this ought to be pardoned. But if all crimes are exceptions, retributive punishment loses its rationale.

Retributivism needs to have two things to be true about human agency, as Frankena suggested in a different context.[41] First, the retributivist, like anyone who has a theory of punishment proper, assumes that the choices and (secondarily) the actions of persons usually have antecedents such as reasons, goals, and intentions. Acts are determined by choices, and the choices in turn have causes. Second, people under normal circumstances are free to act as they choose.

In practice, there are cases in which one or another or both of these assumptions about free will are not satisfied. The examples above are cases in which the second assumption is not satisfied—as when an offender is under the direct or indirect control of another, or is fettered by necessity to a choice between clear evils, or is very, very hungry. The inability of retributivism to justify punishment in cases such as these leaves room for some degree of clemency or full pardon.

14

Justified Pardons: Justifications

Seventy-three-year-old Emily Gilbert was so sick with Alzheimer's disease and osteoporosis that she begged her husband, Roswell Gilbert, to shoot her. He granted her that last wish; he pumped two bullets from a 9-mm Luger into her brain, and then he turned himself in. A jury took only four hours to convict him of first-degree murder; the seventy-five-year-old retired electrical engineer will not be eligible for parole until his hundredth birthday. He was astonished. "I don't feel like I committed a crime," he said. "It's awful. It's the end of my life."[1] He has applied to the Governor of Florida for clemency. Should it be granted?

The moralistic retributivist's answer is yes. Admittedly, it is a moral imperative, on retributive grounds, for a state to punish people who are both liable to punishment and morally deserving of punishment. If the law is a reliable measure of human goodness, so that every crime is a moral failing, then pardon is never justified. But the law and morality can fail to coincide, defeating the moralistic-retributivist justification for punishment. In such cases, pardons are justified. Roswell Gilbert's murder is one such case.

The case illustrates the general justifications for pardon discussed in this chapter, justifications based on the ideas of moralistic retributivism. There are two: pardons are justified when the criminal act is morally right, and pardons are justified when the criminal act, while not morally justified, is evidence that the offender is of good moral character, as is sometimes the case with conscientious acts and acts done in accordance with a set of rules at odds with society's mores.

The Separation of Crime and Sin

Since, according to the theory of pardon suggested here, punishment is required only when there is both legal liability and moral blameworthiness, punishment cannot be required in the case of a person who has broken the law without at the same time acting immorally. Pardons are permissible for a person of whom it can be said, "He broke the law, but he didn't do anything wrong."

Some would argue that a "crime-that-is-not-a-sin" is an empty category. A strong case has been made for the view that, since law is the product of a society with generally held moral views, a body of law incorporates the standards of the people it governs. This may be what Justice Benjamin Cardozo had in mind in *People v. Schmidt*, when he defended the view that all laws define moral duties:

> Knowledge that an act is forbidden by law will in most cases permit the inference of knowledge that, according to the accepted standards of mankind, it is also condemned as an offense against good morals. Obedience to the law is itself a moral duty.[2]

If Cardozo was right, then all offenders deserve punishment in the full retributive sense, and the possibility of pardon will not arise.

However, Cardozo hedged his inference from "forbidden by law" to "an offense against good morals," as he should. The legal prohibition against an act may be good *evidence* that an act is against good morals. But the connection between an illegal act and an immoral act is not a necessary connection. Further, even if one assumes with Cardozo that all laws enact "accepted (moral) standards of mankind," it does not follow necessarily that "obedience to the law is itself a moral duty." It does not follow, that is, unless "mankind" is an infallible judge of moral duty, a premise emphatically and repeatedly refuted by history.

W. G. MacLagan professed to be astonished at the claim that illegal acts are necessarily immoral:

> I find it, I must confess, quite incredible that the publication, by whatever authority, that such-and-such a penalty will be visited on such-and-such an act does, simply by itself, morally justify the infliction of

that penalty. When it is inflicted the victim may, indeed, have no "right" *to be surprised*; but it does not at all follow that he has no right *to protest.*[3]

MacLagan is correct, of course, as even the quickest glance at history will show. However, the issue does not need to be settled here. It is enough to say that *if* someone breaks an immoral law, *then* that person may be pardoned on retributivist grounds.

Pardon for Morally Justified Crimes

Punishment is only required when it is fully deserved because the act was both illegal and immoral. Thus, a retributivist could justify pardoning a person who incites to riot in a country in which revolution is morally required. A retributivist could justify pardoning a person who refuses to answer a draft call to fight in an unjust war, or a person who refuses to comply with state regulations that discriminate unfairly on the basis of race. In all of these example, there is no "sin" for which the criminal "deserves to suffer." And thus the retributivist justification for punishment is defeated.

Many of the proponents of clemency for Vietnam War draft evaders and resisters used just this retributivist argument: Resistance to an unjust war is a moral duty that overrides the duty to obey the law. The Vietnam War was an unjust and illegal war; to quote Norman Cousins, it was "a basic violation of both the natural rights and the legal rights of American citizens . . . illegal in itself."[4] Hence, resisting the war by breaking draft laws was a moral duty, and one should not be punished for doing what is morally right. So the Vietnam protesters should not be punished.

President Ford's Clemency Board did not quarrel with that argument, granting clemency to draft deserters who could demonstrate that they did what they believed was right. For example, they pardoned an applicant who "has been described as a person who is both sincere in his beliefs and of uncompromising moral principle; he repeatedly stated his willingness to go to jail for what he believed to be right."[5]

Mercy killings, such as when Roswell Gilbert shot his wife, though technically punishable as murder, are often pardoned for the retributivist reason that they are not wrong. Records from as far back as 1925 show that pardons were granted to people who murdered for the sake of the 'victim,' to persons who killed a loved one to end his misery.[6] For example, when a father killed his daughter for her sake, it was said,

> The law but does what the law should do, as the code now runs. It cannot concede to the individual, even in such lamentable circumstances as these, the right to take away a human life. To do so would be hazardous in the extreme. And yet, with a force emphasized by this case and its sadness, we perceive that even the law must recognize an occasional exception, though it defeat its own ends to do so.[7]

Consistent with this reasoning, many pardons were issued after Prohibition ended. Reading through the Pardon Attorney's files, I found case after case of liquor law violations that were pardoned. People generally did not believe in the liquor laws and could see no justification for punishing people for doing what was not believed to be wrong.

As in all these examples, moralistic retributivism provides the rationale for pardoning offenders who are liable to punishment but not morally deserving.

Pardon for 'Conscientious Crimes'

It is necessary to be very careful here, because granting pardons for morally justified crimes is a very slippery slope. If some careful distinctions are not made, this theory ends up justifying pardons when they are not really deserved. The situation of *People v. Schmidt* provides an example of the danger. The defendent, Hans Schmidt, had cut up the body of a young woman and thrown the pieces into the East River. He claimed (lyingly as it turned out) that he was obeying God's commands. He told of frequent periods when he felt himself to be in the visible presence of God and under His direct control. When he cut up the woman, he was 'just following orders.'[8] It may be agreed that Schmidt was not morally blameless and did not deserve a pardon on that ground. But how, in theory, can Schmidt be distinguished from Roswell, who also broke the law because he thought it was the right thing to do?

Can a person be blamed and punished for doing what he sincerely believes to be right? Is the most important wrongdoing the breach of some right rule of conduct determined by an independent standard? Or is the retributivist concerned with a moral failure on the part of the agent, a breach of the agent's own sincerely held standards, regardless of their rightness?

Blanshard realized that this was an important problem for the retributivist:

> Retribution is supposed to be based on moral guilt. Very well—how, on that basis, are we to deal with treason? "Lord Haw-Haw" in the last war was a British subject who went over to the Nazis and worked for them actively to the end, when he was caught, tried, and hanged. Probably most of us would find it hard to develop much sympathy for him. Still, it does seem clear that a person might commit treason of this kind for morally high motives, turning against his country with sorrow, but with a conviction that [his country] was bent on a course whose success would be tragic for the world. However mistaken he might be in such a view, he would not be acting immorally in the sense of going against what he thought to be right. Hence, no guilt of the kind required by the retributive theory would be involved in the case, and, so far, he should be set free. The retributionist may say he should still be imprisoned in the interest of public safety. But that is to go over to a different kind of theory.[9]

The problem of conscience is particularly acute for the moralistic retributivist because what has been called the "principle of autonomy" is an important part of retributivist ethics.[10] This principle expresses the view that a moral principle is not binding on a person until he has accepted it because he thinks it is right. While it sounds tautological, it is by no means trivial to say that a person ought to do what he sincerely believes he ought to do, if he has reflected on the situation and holds a moral point of view. Thus the problem arises:

> If, in a difficult case where honest opinions differ, I sincerely believe that you ought to do X and you sincerely believe that you ought to do Y, then what ought you to do? Clearly the principle of autonomy implies that from your viewpoint you ought to do Y. If I also accept the principle I must indeed say that in my view you ought to do X; but in what sense can I *blame* you for doing what, according to the principle, everyone including myself ought to do, namely what they take to be right? Must I not acknowledge that it would be as morally improper for you to defer to my view while you take it to be wrong, as it would be for me to defer to your

view while I take that to be wrong? . . . Though no doubt we normally and rightly assume that the principles which guide our praise and blame are shared by the men we assess, yet in cases of dispute it is theirs and not ours which are decisive; for it is theirs on which, in the sense relevant to praise and blame, they should act.[11]

It is easy enough to say that those who act on the strength of sincerely held convictions are less culpable than those who do what they think is wrong. And it is hard to summon much righteous indignation against people who courageously and steadfastly do what they think is right, even though their moral convictions are minority views and, in fact, against the law.

But this attitude should be sharply distinguished from a passive openness to all honestly held convictions. And it is open to the retributivist, as it is to other moral theorists, to say that a person may as well be led astray by his honest and sincere conscience as by baser feelings. As W. D. Ross wrote,

> If to act in accordance with one's conviction is always, in one sense, to do one's duty, it remains true that one's conscience may be very much mistaken and in need of correction.[12]

Justice Cardozo's response to Schmidt's claim that 'God made him do it,' was sharp and to the point:

> It is not enough, to relieve from criminal liability, that the prisoner is morally depraved. It is not enough that he has views of right and wrong at variance with those that find expression in the law. . . . The anarchist is not at liberty to break the law because he reasons that all government is wrong. The devotee of a religious cult that enjoins polygamy or human sacrifice as a duty is not thereby relieved from responsibility before the law.[13]

R. L. Franklin suggested two counter-examples which are more formal but no less persuasive. He cited two principles that are wrong although they are sincerely held.[14] The first case is that of a person who sincerely holds the following second-order moral principle: It is wrong to reflect on the morality of his first-order moral principles. If his first-order principles are outrageous, no one should hesitate to blame him for having the courage of his convictions. Nor should one congratulate a second sort of person—one who is self-righteous and so confident of his own virtue that he is blind to the faults in his moral principles.

Neither bowing to the dictates of an individual's conscience nor disregarding them in a decision about moral guilt is a satisfactory solution to the problem of conscientious lawbreakers. How can the problem be resolved?

First, it is necessary to draw a distinction between acts that are *morally justified* and acts that are *morally appropriate*.[15] An act that is morally justified is an "act that would be performed by a morally courageous person correctly evaluating all conflicting demands."[16] An act is morally appropriate when it is based on sincerely held moral convictions, correct or incorrect. An act can thus be morally appropriate without being morally justified, although an act cannot be morally justified without being morally appropriate.

The distinction, in other words, is between acts that are *moral* and acts that are *conscientious*. What is important about this distinction is that, while it is the *act* that is praised as 'morally justified,' what is really praised when an act is called 'conscientious' is the character of the *actor* and the connection between the act and the actor's character.

Pardons are justified for moral acts, because the 'offender' has not done anything morally wrong and so does not deserve to be punished. But the issue is more complicated for conscientious, but immoral, acts. The conscientious offender should not be pardoned on the ground that he has done nothing wrong, because he has. But, because he acted conscientiously—showing an admirable character— doubt is raised as to whether he is the sort of person who deserves to be punished. And thus the door is open for a pardon. But only a crack.

It will be remembered that moralistic retributivism seeks an ideal correspondence between wickedness and suffering. So it puts pardoning authorities in the delicate business of weighing the offender's character. On one side is the conscientious motivation for the act; it is praiseworthy to do what one believes is right because it is believed to be right. But this must be balanced against the blameworthiness of accepting false views about what is right, which may be the result of self-righteousness or zealousness or carelessness or even perversity. So, in judging whether or not offenders deserve pardons on retributivist grounds, it is important to know not only what they did and what they believed about what they did, but *why* they held their beliefs—a tricky business indeed.

It would be well to turn to some examples. The first examples are of clearly blameworthy conscientious acts. In the examples that then follow, assigning blame is very difficult. In none of them is an act clearly pardonable, because it is conscientious; however, acting in accordance with principles sincerely held says something important about a person's character, providing information that should be weighed in a pardoning decision.

Unpardonable Conscientious Acts

Consider first an artificial example, based on a suggestion made by Franklin.[17] Suppose a person commits a crime because it is an act required by a first-order principle. Suppose also that the person sincerely holds the second-order principle that it is wrong to reflect on the morality of his first-order principles. In a case like this, there is (literally) no excuse for committing the crime.

Franklin's hypothetical situation was given terrible instantiation by an incident that occurred in a small logging town in Oregon. A mother let her baby starve to death. She would not feed the baby until the baby said grace before its meals, even though the baby was too young to speak, let alone recite the words to a prayer. All the baby could do was cry, which was not good enough. Pressed for an explanation after the baby died, the mother said that she believed that praying was commanded by God *and* that it was wrong to question God's will.

Two pardons that President Reagan granted in 1981 were unjustified on retributivist grounds, because the crimes were based on beliefs that—though sincerely held—were unjustifiably held. Two high-ranking officials of the FBI were convicted of authorizing 'black bag jobs,' illegal break-ins and wiretaps at the homes of acquaintances and relatives of fugitive radicals. The FBI agents were fined a total of $8,500. Their friends in the FBI appealed to Reagan for clemency. Reagan was sympathetic and granted them full and unconditional pardons, because he believed that their convictions grew out of their good faith belief that their actions were necessary to preserve the security interests of their country.[18] They believed, apparently, that lawlessness on their own parts was justified to prevent lawless acts by others.

People who act in accordance with "good faith beliefs" are usually, as Reagan noted, to be praised. But not always. Counterbalancing that view is the blameworthiness of reaching a "good faith belief" carelessly or blindly. The point is that people of good character do not merely do what they believe is right; they also have an additional responsibility as moral agents to come to moral decisions in certain ways. One would not praise Roswell Gilbert if he had flipped a coin to decide whether or not to kill his wife. One does not praise FBI agents who systematically refuse to deliberate about the morality of their acts, believing that their jobs required them "not to ask questions." When a moral judgment is carelessly or blindly made, then the sincerity of the good faith belief is *itself* a character flaw, the mark of a zealot, and such conscientiousness is not a ground for pardon.

Difficult Cases of Conscientious Lawbreaking

There are difficult cases when decisions about pardons have to be based on the narrowest judgments of moral blameworthiness. The political crime—the crime motivated by a political belief—is one. Another is the act that is illegal according to one set of rules (usually positive law) and mandatory according to another set (often cultural).

POLITICAL CRIMES

Pleading for a general amnesty for all the members of the Confederate Army, Union Army General, and later Senator, Carl Schurz said,

> Whatever may be said of the greatness and the heinous character of the crime of rebellion, a single glance at the history of the world and at the practice of other nations will convince you that in all civilized countries the manner of punishment to be visited on those guilty of that crime is . . . almost never [treated] as a question of strict justice. And why is this? . . . Because a broad line of distinction is drawn between a violation of the law in which political opinion is the controlling element (however erroneous, nay, however revolting that opinion may be, and however disastrous the consequences of the act) and those infamous crimes of which moral depravity is the principal ingredient.[19]

He was correct. Unless their punishment is executed during the period of greatest fear and unrest, political offenders are almost always pardoned.[20] All wars in the history of the United States, even minor skirmishes, were followed by pardons for many of those who did not support the war effort or who actively impeded it.

The Vietnam War was no exception. The President's Clemency Board cited as a mitigating factor, "evidence that an applicant acted for conscientious, not manipulative or selfish, reasons."[21] So deserters and draft evaders were pardoned when their religious beliefs required them to refuse military service. But not only that; anyone who had a "sincere, ethical" belief that he should not fight a war, or anyone with "deeply held opposition to the Vietnam War" was a promising candidate for clemency. One applicant was pardoned on the basis of his belief that "peace among human beings is of the ultimate necessity."[22]

The rhetoric accompanying postwar pardons often bubbles over with reconciliation and healing, forward-looking justifications that hold no water for a retributivist. But a retributivist can support a postwar pardon, if it is granted in recognition that the criminal acts were undertaken in accordance with sincerely held political or moral principles about which reasonable people differ.

CONFLICTING RULES

People are bound by many different sets of rules, the criminal law being only one and not necessarily the most important.[23] Family loyalties, religious codes, professional rules of conduct, and customs of an ethnic group control behavior and impose moral obligations. When the criminal law conflicts with a different set of rules, a person's moral obligation is not always clear. Punishing a person who disobeys the criminal law in order to obey his own culture's rules "would be absolutely legal, but scarcely just to them."[24]

The law sometimes takes conflicting obligations into account (as when drug use is permitted during religious ceremonies in certain Native American tribes), but it cannot always do so without undermining its own code. When doing what a culture requires is illegal, punishment imposes suffering on a person who is not sinful, but rather is caught between conflicting moral obligations. So a pardon is sometimes the only morally appropriate response.

Kant himself supplied an example—dueling.[25] For centuries, soldiers (and university students) were honor-bound to accept challenges to duel. But dueling was against the law. Kant argued that the state that persisted in making honor and courage paramount to the soldier could not claim the obligation to punish a soldier who kills another in a duel.[26] In the case of dueling, it is not important only that two codes conflicted; the state actively promoted them both, putting the unfortunate soldiers in no-win situations.

The same is true, Kant argued, of infanticide. If the state fosters the prejudice that an illegitimate child is a disgrace and is outside the law, then the state is being hypocritically self-righteous when it punishes a mother for killing her illegitimate child.

The idea that society should not punish crimes that it encourages recalls to mind Kant's prediction in the *Critique of Pure Reason* that an ideal state will not have crime; this prediction implied that the state's imperfections are at least the occasion for crime—and maybe even the cause of crime.[27]

Alwynne Smart provided a final example:

> Consider two heat-of-the-moment murders where an unfaithful wife is shot in anger by a jealous husband, each crime committed by immigrants from different countries. The two murders are similar *in all respects* except that murderer A comes from a country where a wife's adultery causes a husband and his family great dishonor and humiliation, whereas in murderer B's homeland adultery is regarded as a regrettable lapse, but nothing more. Furthermore, in A's homeland murder in such circumstances is looked upon as a comparatively minor offence, almost excusable. Although both men are equally guilty in the eyes of the law, one is inclined to treat A more leniently than B because he acted under extreme provocation and could not have been expected to view his crime as seriously as we would.[28]

Once the retributivist opens the door to an assessment of moral blameworthiness, it is relevant to consider not just how guilty the offender is in the eyes of the law, but how guilty he is *in his own eyes*.

Conclusion

These are difficult and complex issues. Retributivism does not provide a sorting machine that clanks offenders into the appropriate slot, like dimes and quarters in a parking meter. It does, however, make clear what considerations count in a decision about whether or not to pardon someone who has broken the law. One factor that clearly counts is that the offense was morally justified or, to a lesser extent, morally appropriate.

The lesson to be learned is that a consistent retributivist cannot be strict and self-righteous. Limiting punishments to those that are deserved requires thoughtfulness and questioning. Murphy found this refreshing:

> Such a door, once opened, is hard to close. And once we begin to look at the moral world in this way, we will perhaps be hesitant in being too certain in our judgments as to just what punishment we can in justice demand for others. Once again, here is an insight which leads to moral humility.[29]

15

Justified Pardons: Adjustments to Sentences

When President Gerald Ford pardoned Richard Nixon, he defended his decision by telling the American people that they should take into account how much Nixon had already suffered as a result of his crimes. "I feel," President Ford said, "that Richard Nixon and his loved ones have suffered enough and will continue to suffer no matter what I do, no matter what we, as a great and good nation, can do together."[1] He went on, in a different setting, to question the "propriety of exposing to further punishment and degradation a man who has already paid the unprecedented penalty of relinquishing the highest elective office in the United States."[2] Humorist Russell Baker made fun of Ford's premise by calling it the " 'Hanging is enough, you don't have to draw-and-quarter-em' precedent, a.k.a. the American Tragedy."[3] But with the possible exception of Baker, people generally do take this sort of argument quite seriously. While many did not agree that Nixon had suffered "enough," if he had, they would have accepted this as a *good reason* for a pardon. There is a long tradition of pardoning people whose crimes have struck them down.

Closely related is the tradition of granting pardons to felons who are dying. The story is told of President William Taft, who learned from prison wardens that two prisoners were near death. A cautious man, President Taft sent army and navy surgeons to study the physical conditions of the two men; the surgeons were certain that death was imminent. Thus reassured, President Taft granted them full pardons, so they could go home and die. As Taft later related, "One kept his contract and died. The other recovered at once and seems to be as healthy and active as anyone I know. . . . I was unable to find the evidence of . . . fraud."[4] President Taft may have regretted his decision, but he never doubted the premise on which his decision rested: imminent death is a sufficient justification for pardon.

There is a third related tradition: governors who are opposed to the death penalty in principle commonly commute the sentences of all death-row prisoners.

Finally, presidents most often grant pardons when, years after sentence has been served, the shame and legal disabilities of a criminal record impose hardships that are no longer deserved. A typical case is that of Barbara Lou Ward, convicted in 1963 for embezzling from the Christmas Club Fund of the bank where she worked. Twenty years later, she applied for a pardon, hoping it would solve her continuing employment problems. "I've gone through hell and haven't stopped going through it," she said. When you do something wrong like that, there's no going back. . . . "[5]

These four kinds of justifications share in common the feature that they all look like appeals to pity, but they are not. Instead, they can all be understood as appeals to retributive justice.

Retributive justice proposes that there are specific measures of the amount of suffering criminals deserve to undergo on account of their crimes. Punishment ought to be proportionate to the gravity of the offense, so that the advantage that the criminal has gained is removed. But extraneous circumstances often affect how severe a punishment actually is. So, in at least four different kinds of cases, justice requires some last-minute tinkering with the sentences so that the punishment is no greater than deserved, all morally relevant things considered. The four kinds of cases are those illustrated above: first, when the crime rebounds on the criminal to make the criminal automatically suffer all that is deserved, so that when the debt owed is calculated, the balance is zero; second, when the offender's particular circumstances (old age, mortal illness) make him disproportionately harmed by a punishment, so that the usual punishment would take more than is owed; third, when the set punishment is greater, on an absolute scale, than justice permits; and fourth, when the lingering effects of a felony conviction add punishment beyond what is deserved. Retributive justice justifies pardons in all four kinds of cases.

The Offender Has Already Suffered Enough

Retributive justice provides a measure of how much punishment is required to restore the criminal to an allotment of benefits and burdens that is fair in relation to law-abiding citizens. Thus, however difficult it may be to determine, there is one (or one range of) punishment that is just. If offenders have "automatically" suffered as a result of their crimes, it may be necessary to remit or reduce their punishments so as not to inflict punishments that exceed what they deserve.

Smart offered what may be contemporary philosophy's most hideous example when she spoke of a man who discovers, to his horror and grief, that the victim of the hit-and-run accident for which he is recklessly responsible is his only child, to whom his life has been completely devoted.[6] Smart noted that

> to impose the full penalty would be to impose a total amount of suffering quite out of keeping with the gravity of the offense. There was in this case, a gap between moral justice and legal justice, the possibility of which the law acknowledges when it makes provisions for recommendation of mercy.[7]

An irony should be noted at the outset: because it can justify adjusting a sentence to take prior suffering into account, retributive justice is responsive to human suffering in a way that utilitarian theories of justice can never be. Because retributivists look back, they can see prior suffering. Utilitarians only look forward to judge the deterrent effect of punishment; their philosophical blinders take away hindsight. Hence, the previous suffering of the offender cannot be a mitigating circumstance on utilitarian grounds (unless the offender's prior suffering changes her present or future condition, so that punishment would be expected to affect her more severely than it would have, had she not suffered previously).

In general, only in a retributive theory of justice, where the just punishment is regulated by the offender's desert, can the suffering of the offender be a *direct* reason for pardoning from further punishment. Ironically, the utilitarian system, which primarily weighs pleasure and pain, cannot justify reducing a sentence because the criminal has already suffered enough.

Excusing an offender from further punishment because "the offender has suffered enough" disturbs a nest of problems, philosophical and otherwise. Primary among these is the philosophical problem of defining exactly what kind of suffering counts to reduce punishment.

The clearest kinds of cases in which suffering should reduce punishment are those in which the suffering was *penal*—imposed by the state in response to a crime and *resulting from the crime*. Say, for example, that the appropriate penalty for armed robbery is six years in jail. If a person were arrested for armed robbery and held in a pretrial lockup for three years before being tried and convicted, he could be fairly required to spend only another three years in prison. This much can be granted.

But what if the suffering is a natural consequence of the crime, but is not penal suffering? Nixon's case can serve as an example here; his wrongdoing brought his career crashing down around him. Alternatively, Ezorsky offered a hypothetical example of suffering as a natural consequence of wrongdoing. Suppose that there is somewhere a group of people named Fists. They have a vicious habit of suddenly assaulting innocent people and smashing them with their fists. Fortunately (?), whenever they hit people, Fists suffer terrible pain in their fists—pain so terrible that it exactly equals the amount of pain that should be inflicted on a retributivist measure. Their punishment is the natural result of their criminal acts, and their suffering is, *ex hypothesi*, sufficient. So, Ezorsky wrote, "since Fists receive their full measure of deserved pain through natural causes, they deserve to suffer no more."[8]

Moreover, what of suffering that is not only the natural—but the intended—consequence of a crime: Does it count against the total amount of suffering deserved? Arguably, a self-victimizing crime is a case of ideal retribution, a classic case of "poetic justice"[9] in which the consequences of the crime rebound on the criminal as if the crime had been committed against him. A person who commits suicide is an example; the suicide kills and is killed. There is no slippage here, no gap between theory and practice. This sort of case is a model of justice to one who is a "fundamentalist" believer in the *lex talionis*. The harm done by the crime and the harm done to the criminal appear to be in perfect balance. For the retributivist, is there more to say or to do?

One may go further down this slippery path. Suppose the suffering is neither penal nor the result of the crime. Does that kind of suffering count in the calculus? One may imagine a person whose home washed away in a flood, whose husband left her, whose parents were killed in an automobile crash, and who committed an armed robbery. It must also be imagined that those terrible troubles did not unhinge her or drive her to crime. She no doubt deserves a certain amount of suffering in return for the armed robbery, but has she not already suffered enough? Does her pretrial trouble cancel out her debt to society?

On one view, what may be called Ezorsky's 'Whole Life View' of criminal desert, it does: "the punishment a man should receive for committing a crime at [a given time] can be measurably affected by the unjust sufferings, including those caused by natural disasters, he has undergone prior to [that time]."[10]

This is a path that, once embarked upon, goes a long way. It leads to the conclusion that orphaned rapists deserve less punishment than rapists whose parents are still alive, that a shoplifter could seek a pardon on the grounds that he had been robbed, that the looting that often follows an earthquake is penalty-free. It leads, as Parent pointed out to Professor Ezorsky, to the "necessity of explaining to bemused citizens and incredulous victims alike why on earth pretrial misery per se ought to have any effect on deserved punishment."[11] It might even lead to a pay-in-advance plan, where people could earn rights to penalty-free crimes by serving time in prison before the crime takes place.

Card wondered if this approach also leads to a reverse logic: should the penalty be proportionately *increased* when a person has enjoyed an unusually fortunate life?[12]

> [O]ne may wonder why we do not recognize a principle [requiring that] the penalty deserved for an offense be increased when it is evident that the offender would otherwise, owing to his peculiar good fortune, suffer much less on the whole than he deserves. . . . [13]

These unacceptable conclusions can only be reached if one forgets two important elements of punishment. The first is that retributive punishment is measured not by what offenders deserve, but by what offenders deserve *as a result of their crimes*. It follows that natural suffering unrelated to the crime must not enter into the calculation. This means that any suffering occurring before the crime takes place, and any suffering not directly caused by the crime, is irrelevant to the amount of punishment deserved. Second, punishment must be hardship imposed against the offender's will. So any suffering the offender undergoes by choice—an unsuccessful suicide attempt, for example, or public service undertaken as a sort of voluntary atonement—cannot pay the debt.

These two limitations considerably narrow the scope of legitimate pardons. But the limitations still leave room for pardons when the offender's crime has hurt him as much as he deserves to be hurt as a result of his crime.

Retributive justice provides a rationale for the age-old custom of pardon by *autrefois puni* (literally, 'formerly punished'). Some old criminal codes provided for a pardon in case a criminal was "gravely mutilated or has lost his eyesight" in the course of committing a crime.[14] Likewise, some European codes provide that one "who is sufficiently punished by the wounds suffered in the fight" should not be punished further.[15] If such people were fully punished, on top of this suffering, their punishment would exceed the amount deserved. Where sentencing does not prevent this excess punishment, pardons can be justified.

Moreover, retributive justice provides a basis for criticizing some pardoning practices. For example, three of the mitigating factors used by President Ford's Clemency Board in selecting those deserters and draft evaders who merited amnesty have the appearance of "he has suffered enough" pardons. But each of them fails to meet the limitations defined by principles of retributive justice.

The Board was inclined to pardon draft evaders who had done significant public-service work after their crimes, perhaps on the principle that the public-service work could count as payment of the debt owed. Thus, applicant #150 was pardoned, because

> while applicant was AWOL, he worked as the music director for a number of free concerts and shows which were designed to attract under-privileged youths and to serve as a preventive measure against juvenile crime and drug abuse.[16]

The public-service work might be viewed as compensation for harm done and so count against the amount of *compensation* owed. But punishment is, by definition, suffering imposed against the offender's will, in response to his wrongdoing. Since applicant #150 *chose* to work as a music director, his work there—no matter how much appreciated by the potential juvenile delinquents—was not part of his punishment.

The Board also looked leniently on soldiers who had been injured in the war: "service-connected disability" and "wounds in combat" were two of sixteen mitigating factors.[17] Thus, a soldier who deserted was pardoned, on the grounds that,

> applicant served in Vietnam for one year as an infantryman and grenadier. Applicant was wounded when he found an enemy booby-trapped grenade. He told the men in his platoon to get down, but the grenade exploded in his hands as he attempted to destroy it. He was awarded the Purple Heart.[18]

While the motives of the Clemency Board are understandable and laudable, this ground for pardon is incoherent on retributivist grounds. The terrible injuries were the result of obeying the law, not disobeying it. So, unhappily, those injuries should not enter into the calculation of suffering deserved.

The Clemency Board's reasoning may be based on a confusion, pointed out by Parent, between 'desert credits' and 'desert debts.'[19] Desert credits are what the state owes to the citizen, either in return for special services or because of the citizen's rights as a citizen, Parent explained. Medical care for veterans and equal protection under the law are two examples of desert credits. Desert debts are what the citizen owes to the state: military service, obedience to the law, paying taxes, etc. Parent goes on: "Now the important point about desert debts and desert credits is that they are not thought to be transferrable. Thus a man who wins special recognition for military valor does not thereby earn the right to withhold taxes."[20] Americans owe a debt to Vietnam veterans, particularly those disabled by the war. But the debt cannot be paid in the currency of prison-hours.

'He has suffered enough' is a complicated claim. It is hard enough for a retributivist to specify how much suffering is enough, and having also to determine how much outside suffering counts toward that total compounds the difficulties. Yet, even though there will be difficult cases, there will be cases in which the state's obligation seems clear. The obligation to punish an offender is defeated if punishing him would disadvantage him more on the whole than he deserves to be disadvantaged, considering the suffering he has already been forced to undergo as a result of his crime. Hence, one circumstance that justifies pardon is "recognized consequences more painful than what the law administers as these naturally occur" as the result of the crime.[21]

The Offender Will Suffer Disproportionately

It is clear by now that retributivism makes it very important that people who commit the same wrongs should suffer the same punishment and that those who commit greater wrongs should suffer greater punishments. The usual way to give effect to this dedication to proportionate equality is to impose rigid sets of determinate sentences. But, once again, the individuality of humankind frustrates neatness at every turn. Terminal illness gives a two-year sentence the same effect as a sentence of life imprisonment, and an old man sentenced to a year in prison may expend half of the rest of his life waiting for release. On the other hand, what is a life sentence to a grandfather, when compared to the life a teenage murderer must spend in jail? So many factors must be compared, if comparative justice is to be achieved.

The courts do not usually take age and health into account in assigning punishments, so there may arise a limited number of cases in which the particular circumstances of the prisoner make the usual punishment too severe. These are most likely cases of extreme youth and extreme age and/or illness. In these cases, when the particular circumstances of the offender make the usual punishment undeservedly severe, retributive justice calls for clemency. Some examples follow:

Deathbed Pardons

It is very common for executives to pardon people who are near death. This is often justified as an act arising from pity, so that the prisoner can "go home to die." Governor David Bennett Hill of New York explained this motivation in 1892:

> There is a sort of prevailing notion among the people, or some classes of them, that any prisoner ought not to die in prison, but that he should be released whenever his illness is believed to be fatal. Such people argue that the public interests cannot suffer if the prisoner

should be allowed to die outside of prison walls; and that the dictates of humanity require that himself and his friends should be spared the alleged disgrace of such an ending of his life.[22]

A retributivist cannot justify deathbed pardons as acts of pity, but can subsume them under the principle that people should not be punished more than they deserve. For, by dying, an offender can quickly make a short sentence have the same effect as a sentence of life imprisonment. A pardon prevents such a comparative injustice from taking place.

Pardons Granted for Extreme Youth or Age

The retributivist measure of punishment, again, is this: the offender deserves as much punishment as (and no more punishment than) will remove the advantages unfairly won by an offense. Unfortunately, there is no such thing as a "degree of disadvantaging," as in "this particular rapist deserves three degrees of disadvantaging." Legislators are forced to translate the degrees into the currency of punishment—usually dollars and/or years in prison. The danger is that the retributivist measure gets lost in the translation.

In the case of very old offenders and very young offenders, the danger is particularly real. The standard legislative translations of degrees of disadvantaging presuppose a standard sort of offender. But the seventeen-year-old murderer and the ninety-year-old extortionist are not affected in standard ways by the passage of time. It is hard to sentence them justly without taking into account what percentage of their lives will be taken by the sentence. Again, these calculations cannot be expected from a judge; the responsibility of a judge is to try to achieve a rough sort of justice by following guidelines that lay down standard penalties for standard crimes. But, in extreme cases, when those general guidelines result in sentences that are unjustly harsh, the sentences should be reduced by pardoning officials.

Pardons Based on Particular Vulnerability

In rare cases, an offender pleads for clemency on the grounds that his particular circumstances make him especially vulnerable to punishment, and so the ordinary punishment will, in his case, be extraordinarily (and unjustly) severe. There are cases on record of people who were pardoned to prevent their being deported, a sentence of jail *and* deportation being judged too severe.[23] Watergate defendant John Dean argued, albeit unsuccessfully, that he should not be jailed like a common criminal, because, being an uncommon criminal, he was unusually vulnerable to rape and other harsh treatment from fellow inmates. When such fears are true (as John Dean's turned out not to be), retributivists, with their concern of limiting punishment to what is deserved, must take them into account.

The Punishment Is Just Too Severe

Sometimes, particularly harsh laws require penalties beyond those that can be defended on retributivist grounds. Legislatures lash back against brutal murders, just as frightened voters do, sometimes legislating punishments that have more to do with revenge than with justice. Legislators occasionally appear to be as afraid of appearing "soft on crime" as of being murdered in their sleep. A traditional use of pardons, fully justified on retributivist grounds, is to prevent those excessive punishments from taking effect, thus saving felons from injustice. A secondary effect of using pardons to undercut harsh legislation is to put pressure on legislatures to change the unfairly harsh law by calling attention to its unfairness.

Judges and juries recognize that pardons can provide them a 'way out' when they are forced to impose sentences that even they think are too harsh, and they occasionally count on pardons to correct unjust sentences that they have imposed. The following pardon official was reminiscing about the 1920s, but his words are even more appropriate today:

> I have known a judge who, just after sentencing a man, sat down and wrote our board all the mitigating circumstances in the case while they were fresh in his mind and he live and well, so the convict might have the benefit of it in after years on application for clemency. I have read dozens of communications from judges saying their sentences in specific cases were too severe. . . . District attorneys time and again tell us that particular sentences are excessive and thus confess that a well-intended prosecution was transformed into an unintended persecution. It is a very common thing for us to have petitions for clemency from a majority of the jurors who rendered the verdict of guilty in the given case, and such petitions from all twelve jurors is not a novelty.[24]

In a survey of reasons for presidential pardons between 1885 and 1931, W. H. Humbert found that the recommendation of the judge or prosecuting attorney in the case was *the most common* reason for pardoning.[25] Sometimes it is easier to pardon individuals sentenced under a law than to change the law.

The convict labor system in prisons in the American South provides an example of pardons used to protect prisoners and protest an unfair practice. Believing that the system of forcing convicts to work on chain gangs was operating unfairly, Governor George W. Donaghey of Arkansas pardoned 396 prisoners in one day. The Governor simply decided he could not in good conscience allow such a practice to continue.[26]

In this decade, death penalty and felony murder laws often make governors' consciences squirm. There is no clear retributivist position on the death penalty. Kant, of course, made his position clear: only the death penalty could exact the full debt owed by a murderer. But Richard Lempert recently argued that the retributivist injunction against punishing the innocent rules out capital punishment in a society in which mistakes are made with predictable frequency.[27] The controversy about whether retributivism requires (or permits) the death penalty can be laid aside, however, for this reason: Even a person who supports the death penalty in principle, even Kant, may not be able to support the modern, specific death penalty laws that make death the penalty for a murder that occurs unintentionally during the course of a felony and that make all accomplices to the felony equally liable.

Thus, governors find themselves having to decide whether to allow the execution of a person like Frank Poindexter. Poindexter was a fifty-year-old "baldheaded giant with a limp."[28] In a reference to John Steinbeck's *Of Mice and Men*, Governor Michael DiSalle of Ohio described Poindexter as a "Lenny" who followed a slick con man into an armed robbery and ended up mortally wounding the woman his buddy was robbing. Governor DiSalle finally decided that it was unfair to presume that all parties to a crime resulting in death were equally deserving of the death penalty. He commuted Poindexter's death sentence to prevent the felony murder law from unjustly taking a life.

Other governors have commuted the capital sentences of all death-row inmates in their state[29] because they did not believe that the death penalty was just punishment. They thereby tapped into a long tradition of using the pardon as a protest against what they perceived to be unfair sentencing laws. Because the retributivist position on capital punishment is unclear, it is also unclear whether these actions can be justified on retributivist grounds. But certainly the spirit of retributivism is served by the governors' refusals: punishment in excess of what is deserved cannot be permitted.

Pardons After the Passage of Time

Under present laws, punishments tend to linger long after they should be over. The consequences of a criminal conviction hang around an offender's neck his whole life long. People with criminal records have many sorts of civil disabilities: they cannot serve on juries, they cannot hold government offices, they cannot even lead a Cub Scout pack.[30] And the moral stigma of criminal conviction continues to punish felons after they have served their time.

When the lingering effects of a felony conviction add punishment in excess of what is deserved, then retributive justice requires that a pardon be granted to end this undeserved suffering. In terms of numbers, this is probably the most important pardon condition justified by retributive justice; almost all of the presidential pardons granted today are of this sort. Certainly it is an important function of the Office of the Pardon Attorney to end the ongoing punishment occasioned by the continuing denial of civil rights and moral standing in the community—once the deserved punishment has been inflicted.

Conclusion

Time sometimes plays dirty tricks on philosophers, and one such trick occurred in America in the 1970s. With the best of intentions, retributivists such as the authors of *Doing Justice* pointed out the injustice of individualized sentences based on prospects for rehabilitation. They largely succeeded in substituting the ideal of proportionality for the ideal of rehabilitation and in substituting mandatory sentences for indeterminate ones. They urged that penalties be mild—fewer and shorter jail sentences—and consistently applied. But the "mild" part of their recommendations got lost.

These recommendations for milder penalties were probably ignored because of rising fears and rising crime rates that convinced the public that *something must be done* about crime. Public pressure resulted in harsher laws, in more severe penalties, and, particularly, in the return of the death penalty in many states. The pressure continues, with surveys showing that penalties are still approximately 20 percent more lenient than people would prefer.[31]

The result of the coincidence of these two movements?—harsher penalties that are mandatory. The traditional role of the pardon is to abrogate punishments that are too harsh on an absolute scale, even though harsh punishments are decreed by the law of the land. It is right and proper on retributivist grounds to use pardons to prevent a sentence that imposes hardships beyond those that are deserved. Therefore, it may be anticipated that retributivist principles will define an expanded role for pardons in the next decades, to relieve excessive harshness in late twentieth-century criminal laws and perhaps even to change those laws.

III

Applications, Both Practical and Theoretical

16

How to Distinguish Forgiveness, Mercy, and Pardon

In Part II, I argued that retributivism can define an important role for pardon, as an official act that removes the consequences of a criminal conviction. But what about those qualities that are so often associated with pardon—forgiveness and mercy? Can the principles of retributive justice make sense of them?

The issue has always been a difficult one for retributivism. "Retributive justice" is a comfortable phrase; the two words fit harmoniously together. "Retributive forgiveness" is discordant. It seems silly to look to retributivism for any insight into forgiveness or mercy, because retributivism suffers from such a nasty reputation: "woe-to-him" retribution,[1] atavistic, relentless, vengeful.

For almost a hundred years, retributivism has blushed under Hastings Rashdall's accusation:

> If the duty of punishment is to rest upon an *a priori* deliverance of the moral consciousness which pronounces that, be the consequences what they may, sin must be punished, it is difficult to see how forgiveness can ever be lawful. . . .
>
> It is one of the great embarrassments of the retributive theory that it is unable to give any consistent account of forgiveness.[2]

Rashdall was not alone in his bitter charge. Blanshard also criticized retributive principles for failing to forgive:

Such forgiveness the retributive theory would prohibit. The guilty man must be punished, and, as Kant said, if we remit the penalty, we are guilty ourselves. This seems to me to be an inhumanly doctrinaire rigorism. In the interest of what is conceived to be morality, it would impose a veto on morality at its highest.[3]

W. W. Willoughby was equally critical:

One final proof of the invalidity of the retributive theory may be mentioned, and that is that, when accepted as an absolute principle, no possible room is left for the idea of forgiveness. If it be right that a sin should be punished simply and solely because it is a sin, then forgiveness or remission of punishment can never be other than a violation of that moral law.[4]

In contrast, other philosophers contend that a good reason for adopting retributivism is that retributivism is alone in making logical sense of benevolent responses to crime. For example, Smart said that mercy "is a concept which only makes strict logical sense in a retributivist view of punishment."[5]

On this issue, I am inclined to side with Smart and argue that forgiveness and mercy—as well as pardon—are consistent with a retributive morality. The issue is complex, however, and the stakes are high.

Contrary positions are held so strongly, and critical language is so tart, because the ability to make sense of forgiveness, mercy, and pardon has been used as something of a test case to decide between utilitarian and retributivist positions. An adequate moral theory should be able to provide systematic and principled answers to two sorts of questions regarding forgiveness, mercy, and pardon. First, When are forgiveness, mercy, and pardon moral obligations? Second, What is the meaning of the three related terms?

In this chapter, I briefly outline what I see to be a principal inadequacy of the utilitarian position: its answer to the first question makes it impossible to answer the second. Then I suggest a way that retributivism provides a framework for distinguishing—and justifying—forgiveness, mercy, and pardon.

Utilitarian Grounds for Clemency

Utilitarians have had a hard time making sense of forgiveness, mercy, and pardon. This is true for two sorts of reasons. First, committed as they are to an idealized and universal beneficence, Utilitarians see only one kind of moral reason for doing anything: that the act to be done would, all things considered, do the most good. The only point of punishment could be deterrence or the moral improvement of the offender (or the pleasure gained from vengeance). One chooses severity or lenience, punishment or pardon, accordingly. If a punishment will do more good than harm, it is a moral obligation; on the other hand, if a punishment will do more harm than good, it must not be inflicted. Also, as a practical matter, it behooves institutions of justice to follow rules that tend to have the best consequences on the whole, not in any particular case.

Forgiveness, mercy, and pardon tend to lose their identity in a uniform concern for beneficence. There is no moral difference among them; their only moral content is that they are names for withholding or softening punishment. Rashdall, for example, did not distinguish between forgiveness and remission of punishment.

Moreover, Utilitarianism focuses exclusively on the act and its consequences. Motivational factors such as an attitude of forgiveness, or mercifulness as a virtue of character, or pardon as a performance tied to a special institutional role, do not make an important difference. Thus, while Utilitarianism can provide a standard by which to judge whether or not to punish in a particular class of cases, it blurs the moral distinction between forgiveness and acts of mercy and pardon. Retributivism can do better.

Retributivist Grounds for Forgiveness, Mercy, and Pardon

It is common to see 'forgiveness,' 'mercy,' and 'pardon' used interchangeably. An example is this definition of pardon from Sir Edward Coke, in which all three are invoked:

> A *pardon* is a work of *mercy*, whereby the king . . . *forgiveth* any crime, offence, punishment, etc. . . . .[6]

A similar conflation is characteristic of Utilitarians' criticisms of retributivism's inability to pardon. For example, a confusion of forgiveness with pardon surely underlies Rashdall's accusation that retributivists cannot forgive because they cannot make exceptions to a general practice of punishment. Criticisms like this cannot be answered until mercy, forgiveness, and pardon are distinguished as to their functions and their places in the framework of retributive principles.

Forgiveness

The meaning of forgiveness has been the subject of lively dispute in philosophical journals ever since R. S. Downie reclaimed the topic from the theologians in 1965.[7]

What must first be said is that it is important not to confuse forgiveness with pardon. Granted, several English locutions do encourage that confusion. To "forgive a debt," for example, is to exact less than is due. Similarly, when people ask God to forgive their sins, they are clearly hoping that God will not inflict the full measure of punishment they know they deserve. These people would discover the seriousness of their conceptual confusion if God forgave their sins and punished them nevertheless—which is always an option for God.[8]

It is preferable, I believe, to reserve 'forgiveness' to refer to an attitude of one who has been injured toward the one who has inflicted the injury.[9] The attitude of forgiveness is characterized by the presence of good will or by the lack of personal resentment for the injury. Several consequences follow from this definition.

First, not just anyone can forgive just anyone. "If A forgives B, then A must have been injured [or at least believe herself to have been injured] by B."[10] A person must have, in other words, *standing* to forgive and to be forgiven. It is usurping—often officious usurping—for A to forgive B for injuries to C. This is why howls of protest accompanied Ronald Reagan when he symbolically forgave Nazi wrongs by laying a wreath at a German cemetery that holds the remains of Nazi SS officers: He had no right, not having been personally wronged.

To paraphrase Anthony Flew's remarks about punishment, forgiveness must be *of* an offender *for* an offense against the forgiver.[11] This is true even in two sorts of cases suggested by Downie[12] in which forgiveness seems at first glance to involve a relationship among different sorts of individuals. The first putative counter-example is a person who says that he cannot forgive himself. Such a person may be chastising himself for some injury he inflicted on himself; more often, however, he is refusing to forgive himself for an injury he inflicted on *another*, in which case he may be understood as saying, "My offense is so heinous that it ought not be forgiven by anyone."

In the second anomalous usage, one person says he is unable to forgive another for the injuries done to still another. For example, a father might say, "I cannot forgive her for what she did to my son." Rather than counter-instances, these may be regarded as cases in which the speaker bears a special relationship to the injured such that the injury is also an injury to the speaker. With these two possible exceptions, a first condition for forgiveness is that the person who forgives must have been injured. In addition, the person who is being forgiven must have inflicted the injury.

Second, no particular act or acts need accompany forgiveness. While it is ordinarily expected that forgiveness will bring beneficent treatment, a behavior change is not a necessary—and certainly not a sufficient—condition for forgiveness.[13] Especially important is that no mitigation of punishment need take place.

Forgiveness and pardon are logically independent. A person may forgive a wrongdoer and punish her all the same. For example, if a teenager injures her parents by lying to them and stealing their paychecks to support her drug habit, their resentment may quickly give way to sympathy and concern. They can forgive her, because they love her. Nevertheless, they may bring in the police and have their own daughter prosecuted for theft, in a desperate attempt to change her behavior. Conversely, it is possible to prevent a person's punishment without forgiving the offender. A fearful and abused wife may decline to press charges against her husband while at the same time harboring resentment against him. Likewise, a Governor may grant a full pardon to a rapist without forgiving him, primarily because the governor does not have standing to forgive, not having been raped.[14]

It follows from this definition of forgiveness, that, unlike "I pardon," "I forgive" is not a performative utterance.[15] That is, saying "I forgive" (even under the appropriate circumstances) is not a sufficient and, for that matter, not even a necessary condition for forgiving. Saying "I forgive you" may be like saying "I promise you" in that the listener expects certain characteristic ways of acting to follow the locution; however, unless a change in attitude follows or accompanies the words, no forgiving can be said to have taken place. Alternatively, forgiving can take place without the utterance of any words whatsoever. It can occur, for that matter, without the knowledge, the presence, or even the continued existence[16] of the person who is forgiven.

Third, the most important element of forgiveness is a change in attitude. It is normal to resent an injury. As Murphy argued, the change in attitude called forgiveness is characterized by overcoming or foreswearing resentment toward the wrongdoer—not because the injured person decides the act is not wrong after all, not because the injury is forgotten over time, not for self-serving reasons—but from goodwill toward the wrongdoer.[17]

A. C. Ewing described this attitude. Forgiveness, he said, is a

> right state of feeling and a right mental disposition towards the man who has wronged you, especially in the laying aside of personal resentment. . . . What forgiveness consists in is being as well-disposed to the man who has wronged you as if it had not been *you* that he had wronged.[18]

H. J. N. Horsbrugh suggested a refinement of this view, based on his conviction that forgiveness is, in some sense, a process:

> The decision to forgive is normally only the beginning of a process of forgiveness that may take a considerable time to complete. Indeed there are cases in which it is never completed even though one is committed to completing it once the decision to forgive has been made. For this reason forgiveness may be said normally to have two aspects: (i) a volitional aspect—that which is involved in the decision to forgive, a decision which can usually be partially implemented at once by acting with good-will towards one's injurer; and (ii) an emotional aspect—that which has to do with the extirpation of such negative feelings as those of anger, resentment, and hostility. On the view that I am taking, the process of forgiveness is not completed until one has entirely rid oneself of the sense of injury. The decision to forgive should set this process in motion. But it cannot be carried to fruition immediately through some act of will.[19]

So before people have truly forgiven other people, they must have rid themselves of resentment, or at least they must have begun to eliminate the negative feelings.

A lack of resentment, or an effort to rid oneself of feelings of resentment, must be sharply distinguished from condoning the injury. Forgiveness does not downplay the importance of the injury or pretend that the injury was not a wrong. The opposite is the case. In forgiving an injury, one acknowledges the seriousness of the wrong done. Downie makes this point:

> [C]ondonation is frequently used as a morally inferior substitute for forgiving. In many cases it is easier to play down the extent of the injury and ignore the nature of the moral offence, and so to condone,

than it is to face up to the injury and make the effort to forgive. Moral confusion of this kind will aggravate, and will itself be aggravated by a conceptual confusion between condonation and forgiveness. . . . To forgive is not to condone. . . .[20]

The subjective condition that characterizes the lack of personal resentment toward the offender is a result of goodwill rather than of the insignificance of the wrong.

Fourth, forgiveness is primarily a relationship between persons. Institutions, states, systems of justice cannot forgive—except perhaps metaphorically—because although they may be wronged, they do not resent.

Given this understanding of forgiveness, can a retributivist, who believes that all people who deserve to be punished should be punished, forgive? There is no reason why not, since forgiveness and the duty to punish are not inconsistent. If forgiveness requires a particular attitude toward an offender, it does not necessarily conflict with the action one is morally obliged to take in the case. Thus, nothing in the retributivist position forbids forgiveness.

But is this all? A forgiving spirit is a virtue; surely it is not enough to agree that it is not a vice. Can retributivism provide reasons why one *should* forgive? Yes it can, insofar as a retributivist theory of punishment is justified as an axiom in an overall moral theory. Within such a moral theory, benevolence in the relations among individuals has as prominent a place as does retribution in the relation between legal systems and criminals. Thus, a generally retributivist system such as Judaism, or a retributivist such as Kant, can include a general duty of benevolence and a duty to punish in the same ethical theory.

Consider Kant's view, for example.[21] In addition to perfect duties, Kant defined certain imperfect duties, duties of virtue. "The fulfillment of them is merit . . . but their transgression is not forthwith an offence . . . but merely moral unworth."[22] Among these is the duty of benevolence, which tells people to promote the happiness of others without expecting anything in return.[23] It is not simply the happiness of those who are morally deserving that should be promoted; rather, the duty is to all persons, whether or not they are worthy of love. Insofar as forgiveness is a species of benevolence, there is good reason—though not always compelling reason—to forgive.

Moreover, Kant enjoined his readers to "act as if the maxim of your action were to become through your will a universal law of nature."[24] If it is true that all people will injure another at some point in their lives, and if it is also true that they will want to be forgiven for their wrongs, then, in Murphy's words,

> Is it not then incumbent upon each of us to cultivate the disposition to forgive—not the flabby sentimentality of forgiving any wrong . . . but at least some willingness to be open to the possibility of forgiveness with hope and some trust?[25]

Rashdall's charge that retributivism cannot give a consistent account of the duty of forgiveness misfired. If he really was talking about forgiveness, then he was mistaken. Forgiveness is an attitude possessed by individuals; it is not inconsistent with the obligation of the legal system to punish crimes. However, from the context of Rashdall's argument, it appears that his concern is really with mercy—and *that* is something else again.

Mercy

Showing mercy is an act, not an attitude. Mercy is occasioned by a state of affairs in which A has a legitimate claim—a right or entitlement—against B. Conversely, B is indebted to A; B owes A something. In this, I agree with Murphy, who put forth this definition in his 1986 article, "Mercy and Legal Justice."[26]

A wide variety of relationships could account for this entitlement: A is carrying a loan for B, A is B's landlord, A has won a civil suit against B, A was promised something by B, A is the father of B, and so forth. Mercy is shown when A decides not to exercise that right—not to exact payment—or to exact less than is due. Only the holders of the entitlement can relinquish it; no one can do it for them.

It follows that a person who is treated mercifully has no legitimate claim to the merciful treatment. Moreover, the relinquishment is merciful only when A acts out of pity for B—out of a desire to lessen B's suffering. Any other motivation changes the nature of the act. This much is evident from a comparison among four alternative sets of parents, each of whom loaned their daughter enough money to buy her first house, the money to be repaid in ten years:

> After ten years, the first parents say, "Without pay, our daughter has cared for us in our illnesses for ten long years. She owes us nothing. We cannot ask her to repay the loan."

That is fair, but it is not merciful.

> The second parents say, "With all our retirement income, we really do not need the money. Let's not ask her to repay the loan."

That is generous, but again, it does not have the quality of mercy.

> The third parents say, "Let's not ask her to repay the loan. Maybe she will be so grateful that she will leave Chicago and come live near us."

That is cagey, but it is not merciful.

> The fourth parents say, "If we ask for repayment of the money now, she will never be able to pay all those doctor bills. Let's forget the loan."

That, finally, is merciful.[27]

H. R. T. Roberts offered this formal definition of mercy: "In all justice, X owes me A (I am entitled to A from X), but it is mine to exact and I choose not to."[28] He suggested that the following cases qualify as instances of genuine mercy:

> Suppose X was due to pay you a sum of money on a particular date, but told you that he would be greatly inconvenienced to have to do so. If you could easily enforce your claim, but decide out of compassion to give him more time, you could be said to be acting mercifully. Again, suppose you were driving along and another driver hits your car through what is clearly established as his fault, and you say, 'Well, I suppose you'll have a rough time if I make a case of it, so I'll let it go this once,' your act could be called one of mercy. Suppose, once more, that you discovered that your wife had been unfaithful and that you could ruin her by a divorce, but for her sake decided to give your marriage another chance, you could be said to exercise mercy.[29]

All three cases illustrate a species of the virtue of compassion—deciding out of pity to exact less than the full amount to which one is entitled. They are typical cases of mercy in that the leniency is gratuitous.

Is mercy, so defined, permissible on retributivist grounds? That depends on who or what is acting mercifully.

Among *individuals*, nothing in the retributivist's view forbids a person from deciding not to exact what is due. In fact, the same principle of benevolence that makes forgiveness a virtue also would recommend showing mercy. For, in relinquishing a right in order to relieve another's suffering, one promotes the happiness of others without expecting anything in return, as the principle advises.

But that is for individuals. The story is much different for *laws and institutions*, which cannot be merciful for logical reasons. Since mercy is a virtue of persons as individuals, a species of compassion or leniency, a specific moral response from one individual to another, a law cannot be merciful, nor can institutions. A law cannot be merciful because it cannot both establish and relinquish a right. Suppose a law says, "Any person who drives faster than 55 mph will be fined $50, except those in pitiable circumstances, who will be scolded." This law is not itself merciful, because it does not relinquish a right or exact less than the amount due for reasons of pity. What it does is make a more complicated statement of what is due. An institution cannot be merciful for the same reason. The flexibility built into its systems may vary that to which it is entitled; but if the rules or practices require than an entitlement be relinquished, then there was no entitlement to *be* relinquished in the first place.

Now, what of *judges*? It is by now clear that, in their personal lives, they should act mercifully. But may they act mercifully *in an official capacity*?

Since the 1960s, philosophers of a generally retributivist bent have tried to understand what mercy is in a judge and how it is to be justified. The issues have been particularly intractable.

Smart may be credited with starting it all by asking, What are the conditions under which the exercise of mercy is appropriate, when mercy is understood as the act of a judge who exacts less than the just penalty for a crime?[30] Her answer: Precious few. With some specifiable exceptions, showing mercy is either unjust or it is not truly mercy.

Mercy in judges is unjust, Smart argued, because when judges act mercifully, they have already decided on the deserved—and thus proper and just—punishment. If they choose not to inflict the full measure of punishment, by acting mercifully they are failing to do what is just; hence, they are acting unjustly. Moreover, merciful judges violate the principle that like cases should be treated in a like manner and so commit an injustice relative to other wrongdoers.

On the other hand, Smart continued, sometimes when judges reduce or eliminate penalties, they do so in order to adjust the penalty to the degree of suffering a wrongdoer really deserves. In these cases, because the mercy is not granted from benevolence, but from a sense of justice, it is not really mercy, but 'mercy,'[31] legal justice,[32] or pseudomercy.[33]

What of the few exceptions? Mercy is justified and is truly merciful, Smart wrote, only when a judge is compelled to mitigate a punishment by the overriding claims of other moral obligations, as, for example, the obligation to prevent undeserved suffering on the part of the offender's family. Thus, mercy can only be justified—and then only in a few cases—by a multidimensional moral theory. How this avoids comparative injustice, Smart did not say. Nor did she explain how the judge avoids usurping the rights of the people.

Card was not so willing to introduce conflicting moral obligations; the obligations of justice are quite enough to justify clemency, she believed. Punishment and mercy are both just responses to the offender's desert, Card argued. Legal justice concerns what the offender deserves as a result of wrongdoing. Moral or "cosmic"[34] justice measures desert by an offender's character and particular misfortunes as well. Card formulated cosmic justice in a desert principle:

> A person deserves to suffer, on the whole, no more than one could reasonably be expected to suffer from . . . others were he to live in a community in which everyone else had the same . . . sort of over-all character as he.[35]

Mercy is a virtue shown by judges, Card said, when they mitigate a punishment to insure that the legally permissible punishment does not exceed the amount of suffering the wrongdoer deserves, all things considered.

It may clarify Card's view to point out its similarity to Aristotle's view of the relation between two kinds of justice—equity and legal justice. Aristotle explained that,

> the equitable is just, but not the legally just but a correction of legal justice. The reason is that all law is universal but about some things it is not possible to make a universal statement which shall be correct . . . When a law speaks universally, then, and a case arises on it which is not covered by the universal statement, then it is right, where the legislator fails us and has erred by over-simplicity, to correct the omission.[36]

In much the same way, Card understood mercy as an expression of justice, a perfection of justice. In contrast to Smart, she argued that tempering justice with mercy is not being a little unjust, but, rather, is being even more just (than legal justice).

This leads to what I would call 'the Dilemma of the Merciful Judge.' Roberts was quick to point out the problem[37]: When an offender is treated mercifully, either the offender *is* given the penalty deserved (in which case the offender is not being shown mercy, but justice) or the offender is *not* given the penalty deserved (in which case the offender is being treated unjustly). Thus mercy is either justice or it is unjust.

In cases in which legal and moral justice are at odds, Card's view also leads to what may be called 'the Dilemma of the Unjust Judge': If judges impose legal justice, then they are being morally unjust. If judges follow the dictates of moral justice, then they are being legally unjust. Thus, when legal and moral justice are at odds, the judge must always act unjustly.

In addition, if all of the above is not enough to persuade a judge not to be merciful, there is this final objection to judicial mercy: If judges decide to override considerations of justice and act out of pity, they are usurpers. They are relinquishing a right—the legitimate expectation that offenders will be punished—that belongs not to them, but to the people.

A judge cannot be merciful, because the debt owed is not the judge's to exact. The offender owes the judge nothing (or nothing more than is owed to every other law-abiding citizen). And if there arose the rare instance in which a judge might be in a position to sentence a criminal who has injured her personally, the judge would be obliged to disqualify herself to avoid another kind of injustice altogether, the injustice of being a judge in one's own case. Even when the relationship between God and miserable sinners is taken to be the model for mercy, it cannot be God the judge who is merciful, but God the father.

So I would argue that a judge cannot exercise either pseudomercy or real mercy in an official capacity. But individuals can, and indeed should, exercise real mercy in their relations with other individuals.

Still unaddressed are the questions of whether, leaving aside issues of mercy and forgiveness, the official remission of punishment can be justified. If so, by whom? And through what device? The answers, as detailed in Part II, are: Yes, by the executive, through the pardon.

Pardon

It should be clear from the preceding discussion of forgiveness and mercy that I do not take pardon to be closely connected with either forgiveness or mercy. This view of pardon repudiates an historically respected view. Chief Justice John Marshall gave judicial recognition to the view that a pardon is an act of mercy when he wrote for the court in *U.S. v. Wilson*, defining a pardon as "an act of grace proceeding from the power entrusted with the execution of the laws."[38] This correlation was abandoned by the Supreme Court in 1927 when Justice Holmes wrote for the court in *Biddle v. Perovich* that a pardon is "not a private act of grace from an individual happening to possess power,"[39] but an act for the public welfare.

This book follows this more positivistic direction by saying that an act may be characterized as a pardon if it has the following characteristics. A pardon is an act by the executive (or others legally empowered) that lessens or eliminates a punishment determined by a court of law, or that changes the punishment in a way usually regarded as mitigating. A pardon is an act one can perform only in a social or legal role. This characteristic distinguishes it from forgiveness and mercy, which are virtues that persons exhibit as individuals.[40] Anyone who has been injured can forgive, but only one formally constituted within a legal system is qualified to pardon a violation of the norms of that system.[41]

In part because of the institution-bound nature of a pardon, pardoning is normally performative, as forgiveness and mercy are not. That is, when uttered by an appropriate person in an appropriate setting, "I pardon you" constitutes a pardon. For example, when Gerald Ford said,

> Now therefore I, Gerald R. Ford, President of the United States, pursuant to the pardon power conferred upon me by Article II, Section 2, of the Constitution, have granted and by these presents do grant a full, free, and absolute pardon unto Richard Nixon . . . ,[42]

he set in motion the legal machinery to insure that any offenses would not be punished.

According to retributivist principles, pardons are justified by a reassessment of the moral guilt of the offender. So a pardon neither condones nor overlooks crime. It acknowledges a mistaken presumption: the legal system presumed that anyone who did that sort of thing would deserve to be punished, but the special facts of that case overrode that presumption. Consequently, a pardon does not imply guilt.

In this regard, the view of pardon held here conflicts with the predominant view of pardon in U.S. courts, although there is still some judicial debate about whether or not a pardon implies guilt.

For example, after the Nixon pardon, scholarly tempers flared over the issue of whether or not President Ford's action implied that Nixon had committed crimes for which he would have needed to be pardoned in the future. Nixon's representatives adamantly denied that being pardoned requires any wrongdoing to have been done. Ford's spokesmen were not as sure. And legal authorities could not agree among themselves. The court's dictum in *Burdick v. U.S.* suggested that a pardon does imply guilt:

> This brings us to the difference between legislative immunity and pardon. They [sic] are substantial. The latter carries an imputation of guilt; acceptance a confession of it. The former has no such imputation or confession. . . . It is the unobtrusive act of the law given protection against a sinister use of his testimony [sic], not like a pardon, requiring him to confess his guilt in order to avoid a conviction of it.[43]

Justice Learned Hand was more explicit:

> It is suggested that a pardon may not issue where the person pardoned has not at least admitted his crime. I need not consider this, because everyone agrees, I believe, that if accepted the acceptance is at least admission enough. It is an admission that the grantee thinks it useful to him, which can only be in case he is in possible jeopardy, and hardly leaves him in position thereafter to assert its invalidity for lack of admission.[44]

In answer to Justice Hand, it must be noted that a pardon might be very useful to an innocent person who is in jeopardy of being unfairly punished. It is not inconceivable that a person might accept a pardon for an offense of which the person is innocent, simply to avoid the costs in time and trauma of a criminal proceeding and the risks of false imprisonment.[45] So a person could accept a pardon—and thereby admit it is useful to him—without admitting guilt. It is possible, in other words, for a person to accept a pardon for a crime he never committed.

Authorities, both legal and philosophical, do not always agree with Justice Hand. An example is Downie:

> To pardon a person—whether this be done by monarch or club com-
> mittee—is to let him off the merited consequences of his actions; it is
> to overlook what he has done and to treat him with indulgence. To
> pardon is in fact to condone.[46]

On the contrary, in the view I have been defending here, pardons
imply not guilt, but legal or moral innocence or both.

If so, what should be the consequences or legal effect of a
pardon? This is a perennial question in legal theory. One would think
this would be an easy one: a pardon should wipe out guilt and punishment. And indeed, there is ample precedent for this view. Justice
Stephen Johnson Field described the results of a pardon in *Ex parte
Garland*:

> A pardon reaches both the punishment prescribed for the offense and
> the guilt of the offender; and when the pardon is full, it releases the
> punishment and blots out of existence the guilt, so that in the eye of
> the law the offender is as innocent as if he had never committed the
> offense. If granted before conviction, it prevents any of the penalties
> and disabilities, consequent upon conviction, from attaching; if granted
> after conviction, it removes the penalties and disabilities and restores
> him to all his civil rights; it makes him, as it were, a new man, and
> gives him a new credit and capacity.[47]

Alas, the issue is not so simple, for convictions leave long and
subtle trails. Can a pardoned felon, arrested again for a similar crime,
be sentenced under repeat offender statutes? Is a pardoned felon a
person "of good moral character" for purposes of immigration and
gaining membership in the bar? Does a pardon restore the civil rights
of the offender—the right to hold public office, the right to serve on
a jury, and so forth?

The questions are complex, the literature prolix,[48] and the issues
far from settled. The view of pardon adopted here suggests a resolution: If a pardon is granted because there is no legal guilt, then all
legal consequences of conviction should be removed including jail
term, fine, and probation. If a pardon is granted because there is no
moral guilt, then all consequences bearing on offenders' moral standing should be removed; that is, they are not repeat offenders, they
are of good moral character, and they should regain their civil rights.
If a pardon is granted because there is neither legal nor moral guilt,
then all consequences whatsoever should be removed.

In summary, a pardon is an act of the executive that mitigates or entirely remits the punishment of an offender and restores the offender to innocence in the eyes of the law.

A pardoning executive is not giving up a strict insistence on justice in favor of a (supposedly higher) good, namely benevolence. If this were the only genuine case of pardon, then pardon could not be justified in a system of retributive justice. Its critics would be correct that retributive justice makes no room for pardon. However, pardon is justified in retributive justice as a way to bring an offender's liability in line with moral desert, to make sure that the offender gets "just deserts," no more and no less.

Conclusion

A reexamination of the concepts of forgiveness, mercy, and pardon shows that critics who accuse retributivism of being rigid, unforgiving, and self-righteous are too strict themselves.

Retributivism does not forbid forgiveness. On the contrary, forgiveness, an attitude of goodwill toward a transgressor, is always possible for a retributivist. Retributive justice requires punishment under specified circumstances, but it makes no restriction on how people must feel toward someone who wrongs them.

A retributivist also can be merciful. People act mercifully when they decide not to exact the full measure owed to them by another. Retributive principles do not forbid mercy in the relations among individuals, although genuine mercy is out of place in the institutions of a legal system.

Finally, retributivism gives meaning to the concept of pardon and makes an important place for it in a system of punishment. The purely benevolent remission of a truly just punishment cannot be justified on retributivist grounds alone. But retributivism permits pardons when they are necessary to bring about a more perfect form of justice.[49]

The pardoning power is a prerogative that allows the executive to soften the strict harshness of the law by determining a sentence that takes into account factors that have a moral significance—though they may lack a legal one.[50] While this flexibility cannot be expected from the laws themselves, more can be expected from those who administer the law.[51] The legal tool they use is the pardon.

17

How to Abuse the Pardoning Power

When a person's reputation is ruined, it is common to say, "His name is mud." The expression comes from a story that began the night President Abraham Lincoln was shot. His assassin, John Wilkes Booth, made his escape by jumping from the President's theater box to the stage, breaking his leg. In serious need of medical attention, Wilkes covered his famous face with a false beard, rode through the night to a country village, and knocked on the door of Dr. Samuel Mudd. Dr. Mudd set the stranger's leg and thought no more about the nighttime call until he learned the next day of the President's assassination.

His suspicions aroused, Dr. Mudd went straight to the police, who arrested him as a conspirator in Lincoln's assassination. He was treated cruelly and unfairly in jail and at trial—chained, his head tied into a canvas bag, his ears stuffed with cotton, denied an attorney, his witnesses intimidated. He was convicted and sentenced to life in prison. His name was Mudd, dirtier than dirt, trampled underfoot.[1]

Four years later, a yellow fever epidemic swept through the prison where Dr. Mudd was held. Dr. Mudd fought the epidemic with skill, selflessness, and tireless courage. When it was over, Dr. Mudd was pardoned, "in recognition of his heroism in fighting the epidemic."[2] Since the pardon made no reference to his innocence or to the injustice of his sentence, his name was still Mudd.

Was that an abuse of pardon?

It might be argued that the question is perverse. The pardon had the good effect of releasing from prison a man who probably should not have been there in the first place. So why quibble with the justification? After all, the use of the pardon is not reviewable, and so criticism (short of impeachment) can have no effect and is a waste of time and rancor. Also, does not the world always have room for kind rewards and good results?

Why Pardons Should Be Criticized

It is true that the courts do not have the standing to overturn or even, some would argue, to criticize a pardon. When courts comment on acts of clemency, it is usually only to reaffirm that, in this matter, the executive can do just as he pleases. "The Constitution clothes him with power to grant pardons, and this power is beyond the control or even legitimate criticism of the judiciary," declared the Texas Court of Criminal Appeals.[3] Chief Justice William Howard Taft pointed out that, "Whoever is to make [the pardoning power] useful must have full discretion to exercise it. Our Constitution confers this discretion on the highest officer in the nation in confidence that he will not abuse it."[4]

But that no one can make a President change his mind does not mean that a President cannot make up his mind wrongly. Moreover, that his decisions are unreviewable makes it even more important that they be subjected to criticism, for this reason: Criticism makes it clear that there are standards to be upheld and makes clear what those standards are. Criticism reaffirms that pardons should not be given out on whim, capriciously. Pardons should be granted as scrupulously as punishments are imposed—and, as I have argued, according to the same principles of desert that limit the assignment of punishments.

An analogy may make my point more emphatically: In most universities, academic freedom extends to the assignment of grades. Usually, no one can force a professor to change a grade (particularly a good grade). But a professor who assigns grades for unjust reasons (in return for favors done, or out of pity, or to promote someone in a position to help the university) is subject to criticism. The criticism performs the important role of defining and reaffirming the proper standards. And, if the abuse of the grading power is serious enough, the professor probably will be fired—on grounds of moral turpitude. Just so, a President who gives out pardons for the wrong reasons should be criticized and—if the injustice is outrageous—impeached.

Also, may the idea that there can never be too much of a good thing, so often used to protect pardons from criticism, be finally laid to rest? Of course there can be too much of a 'good thing,' as Aristotle pointed out in regard to the virtues, and as Bentham and Kant (and about everyone else) pointed out in regard to pardons. If an example is needed, one is provided by the hapless members of the Mudd family who were still trying to clear their family name more than a hundred years after their greatgrandfather went to jail.

So, it is important to become clear—through criticism—about what kinds of reasons are bad reasons for granting pardons. Specifically, in what sorts of cases is the pardoning power abused?

Improper Uses of the Pardoning Power

The retributivist's explanation of what constitutes an abuse of the pardoning power could be quite short: the pardoning power is abused when a pardon is granted for any reason *other than* that punishment is undeserved. But that is an unhelpful answer; it is far better to identify the general types of cases in which pardons cannot be justified on retributivist grounds alone.

It should be recognized throughout that a given pardon can be justified by a number of different reasons, so a failure to be justified by one sort of reason does not necessarily mean that a pardon is not justified at all.

Pardons for the Public Welfare

The most frequently granted undeserved pardons are those granted only because it is in the public interest to grant them. Sometimes these pardons are actually granted against the offender's will; even those granted with the offender's consent may be coercive, given the forced choice between unwelcome alternatives. In this category are pardons granted to induce the offender to do something: to turn state's evidence, to rejoin the army, to populate the colonies, to testify regarding the details of the offender's own crime, to volunteer for a medical experiment, and so forth. In 1730, for example, a pardon was granted to a condemned criminal on condition that he let a physician cut a hole in his eardrum to study its effects on his hearing.[5]

Pardons like these cannot be justified on retributivist grounds. It is as unjust to *pardon* an undeserving person only for the public welfare as it is unjust to *punish* an undeserving person only for the public good. In each case, the offender's own deserts are overridden, and the offender is used against his will as a means—a tool—for the achievement of the state's ends.

Pardons granted for the public good are not necessarily beneficent from the offender's point of view; sometimes they are coercive, and unjustly so. The case of *Burdick v. U.S.*[6] illustrates how a pardon can be used against an offender. George Burdick was the city editor of the New York *Tribune*. Called before a federal grand jury investigating custom fraud, Burdick declined to answer questions about the sources of his stories, citing his right against self-incrimination. President Woodrow Wilson granted him a pardon for any crime he might be questioned about by the grand jury, which called him again. Since the pardon would have effectively deprived Burdick of his Fifth Amendment rights, he refused to accept it. The Supreme Court went along with Burdick this time.

But it only took them twelve years to reverse themselves, deciding in *Biddle v. Perovich*[7] that acceptance is irrelevant to the validity of a pardon. If, in the President's judgment, pardon serves the public interest, pardon can be granted (imposed?) against offenders' interests and against their wills. Accordingly, pardons have been used to limit constitutional rights, to deliver an offender into the hands of another state for a more severe punishment, to facilitate deportation, and to override an offender's decision that execution is preferable to life in prison. How can this be justified? Because, said the court, a pardon is justified solely by the executive's decision that the public welfare will be better served by not punishing than by punishing.[8]

The retributivist must directly oppose this sort of forward-looking, utilitarian calculation in the assignment of pardons, as in the assignment of punishments. The executive's decision is not enough; it must be the correct decision. And even then, considerations having to do with the offender's rights as a rational agent and the seriousness of the crime make it clear that it is not always just to do what is good. The public welfare does not automatically override the individual's right to be treated as he deserves.

Pardons justified by appeal to the public good often display a second sort of injustice—the comparative injustice of treating similar offenders in different ways. A dramatic illustration came from the old town of Regensburg on the Danube.[9] The year was 1386, and the town was threatened by the armies of the Dukes of Bavaria. To prepare for war, all the people who had been banished from the town were pardoned and asked to return; all, that is, but women and church burglars. Since women (and presumably church burglars) were not useful to the state's war effort, their banishment continued.[10]

Pardons granted on days of celebration—Christmas, Thanksgiving, Good Friday, the birthday of the Governor of Oklahoma[11]— might seem to make no sense at all. But if pardons increase the popularity of the ruler, and if a popular ruler is in the public interest, then holiday pardons are in the public interest, and an attempt to justify them may be made on that ground. Governor James Withycombe of Oregon put his finger on the retributive objection to this practice: "If a man is entitled to a pardon, he is entitled to it regardless of whether or not it is due him during the holiday season. If he is not entitled to it, the fact that it is the holiday season is no reason why leniency should be extended."[12]

And if this retributive objection is not enough to put holiday pardons on the list of abuses, Governor Friend William Richardson of California offered a pragmatic consideration: "Californians will, I believe, enjoy this sacred day better with the knowledge that a score of murderers, robbers, and pickpockets has not been turned loose upon them today."[13]

The objection to public-good pardons leads retributivists down an increasingly rocky road. The retributivist argument is relatively easy to make for the preceding examples, because those examples fit with intuitions about unfair treatment. But the retributivist who believes that moral desert is the only ground for justified pardon will have to object on the same grounds to some pardons that are intuitively right and good. These are pardons that benefit society while they also benefit the offender. For example, a retributivist is forced to take issue with Alexander Hamilton and protest that a "well-timed offer of pardon to quell an insurrection" is unjust, if it is undeserved on moral grounds.

There is usually some room to maneuver, though. Pardons to end rebellions are not justified *on that ground alone*; however, they might be justified by a reassessment of the ill-desert of the rebels. This seems to be the case with the Lincoln and Johnson pardons of

Confederate soldiers, which were justified for mixed reasons: to promote harmony (a justification that by itself cannot be admitted by a retributivist) and to acknowledge that the Confederate soldiers were not really traitors (a most retributive justification).[14]

Recent events in Argentina provided a severe test of the retributivist insistence that pardons "for the public good" cannot be justified on that ground alone. Military officers who had run the country for years had committed dreadful human rights violations. The new government, determined to bring them to trial, faced a revolt of the armed forces who demanded that the officers be pardoned. To save itself, the government passed legislation to pardon most of the officers, on the express grounds that most of the soldiers facing charges for murder, torture, or other human rights violations acted "under orders during a state of emergency," and so were not blameworthy.

It is very difficult to tell a government that it is unjust to take the steps necessary to save itself. Would even Kant expect a just society to forbear from the one unjust act that would keep an unjust revolution from sweeping the country? Fortunately, what cannot be justified on one ground can often be justified on others. Because there is a substantial social cost when unjust pardons are granted, it is more the rule than the exception that a pardon that *truly* serves the public interest is a pardon that is deserved.

Pardons to Promote the Private Welfare of the Pardoner

This abuse of pardon is corruption, pure and simple—using a pardon for the selfish ends of the pardoning executive. It almost goes without saying that pardons granted for only selfish reasons cannot be justified; what perhaps needs some explanation is how deeply this particular knife cuts.

Pardons were a reliable source of income for many a monarch, with England's James II probably setting the record for the most profit made from one pardon (16,000 pounds sterling). American executives are piffling in comparison, with the American record-holder, Oklahoma Governor J. C. Walton, a far distant second to James II. Governor Walton is the only American executive to have been impeached for abuse of the pardoning power.[15] The article of impeachment said, in part,

> the said J. C. Walton knowing at the time that these close friends and
> special officers were representing to the friends and family of the party

> desiring a pardon that by virtue of their influence with said J. C. Walton that they could and would obtain a pardon or parole for the party desiring it, upon the payment to these friends or agents of a large sum of money; that said J. C. Walton delivered numerous pardons and paroles granted to persons accused of crimes wholly unworthy of clemency. . . . [16]

Nobody thought that the said J. C. Walton was doing this for his *friends*. Said Senator Darnell, explaining his vote: "I cannot bring myself to believe that Jack Walton, who . . . has clearly prospered financially far beyond the possibility of his meager salary, and whose signature was absolutely necessary before the stream of filthy lucre began flowing, permitted all of the money to go to his friends and refused to participate in the distribution of same."[17]

No one will defend pardons for bribes. But what about pardons granted for political ends? Montesquieu pointed out that a pardon can increase the popularity of a monarch, "so many are the advantages which monarchs gain by clemency, so greatly does it raise their fame, and endear them to their subjects. . . ."[18] So much greater are the advantages to an American executive whose livelihood depends on being elected, that it is a wonder more use is not made of pardons. Certainly, history provides one example of the political misuse of pardons: Governor Coleman Blease of South Carolina, who promised that he would pardon the crowd's favorite, if the crowd would deliver the election.[19] Governor Blease defended his action with a notably repulsive argument:

> I took the position that I was the servant of the people . . . and when a community where a crime had been committed, with the best people, the white people, signing the petition, said that the criminal had been punished enough, I turned him out without regard to criticism.[20]

Not many governors are like Governor Blease. And, although "mention is often made of commutations of minority-group members on the eve of important elections,"[21] there is no evidence to suggest that "any decision has been motivated solely by the political exigencies of the moment," as Abramowitz and Paget delicately concluded in their 1964 study of executive clemency.[22]

It is not overly cynical to suggest, however, that politicians weigh public opinion in a pardon decision. President Lyndon Johnson was so disgusted by press criticism of the large number of pardons he granted in one year that a following year he refused to grant a single pardon.[23] For President Ford, on the other hand, his pardon of Richard Nixon may well have cost the election, an outcome he may have anticipated and disregarded. Sometimes it is hard to tell what effect public opinion will have on pardon officials, who may be either offended or intimidated by publicity campaigns and petitions overflowing with signatures. And it is hard for politicians to gauge public reaction to a pardon.

Public opinion is important when it comes to making the laws that define crimes and punishments. But pardon decisions counteract the effect of the laws, make exceptions to the laws. Public opinion should not be heard here. Pardoning to win an election is morally indistinguishable from pardoning to win a buck.

Pardons to Reward Past Actions

What should the retributivist position be in regard to pardons like Dr. Mudd's—pardons granted to reward the pardoned person for having done something for the state? In this category are pardons granted as rewards for acts done in prison (quelling a prison disturbance, fighting an epidemic in jail, volunteering for a medical experiment), as well as rewards for acts done before the person went to jail (having been a good Governor, having fought with valor in the Vietnam War, having been a President who dedicated his life to public service).

Pardons as rewards cannot be justified on retributivist grounds alone. Society and governments have many appropriate ways to reward people for their services. Pardon is not one of these. To reward an offender with a pardon brings about the injustice of punishing a person less than he deserves and the comparative injustice of punishing two similar offenders differently, both violations of retributivist principles of justice. To the extent that these pardons are not connected with what the offender deserves *as a result of his offense*, they cannot be justified by retributivist standards.

But this proves to be a difficult line to draw, for it is possible for a retributivist to argue that a person's past sacrifices should be somehow calculated into the total amount of suffering that the offender deserves.

When President Ford's Clemency Board reviewed cases to decide which draft deserters should be granted amnesty, one of the factors taken to mitigate guilt was "service-connected disability."[24] Deserters were more likely to be pardoned if they had had experiences like these:

> During one of applicant's combat missions, a hostile mine explosion caused him to suffer leg and ear injuries. As a result of his hearing loss, he was restricted from assignments involving loud noises.

> Applicant was medically evacuated from Vietnam because of malaria and an acute drug-induced brain syndrome. Since his discharge, he has been either institutionalized or under constant psychiatric supervision.[25]

These people already paid their debt to society, the Clemency Board decided. It was only fair to deduct their prior suffering from the total suffering they deserved for having deserted. Such pardons are not exactly rewards for services rendered, the Board decided, but recalculations of the debt owed.

This is an ill-advised argument, however. Punishment is not just *any* suffering; punishment is suffering imposed by the state in response to a crime. Society should do its best to pay its debt to the soldiers who were injured in the war (certainly, far more than society has done, which is scandalously little), but pardon is not the appropriate currency.

Pardon for Pity's Sake

Retributivists are often reluctant to say so, for fear of being accused of heartlessness, but in pardoning decisions, "mawkish sentimentality, and the tears of mothers and wives, must alike be disregarded."

Yes, prisoners suffer in prison, and it is not until one begins to study pardon petitions that one realizes *just how much* prisoners suffer in prison. "No matter how firm a stand he may appear to take, no matter what he may think and argue to the contrary," said Oregon Governor Oswald West,

> no man with a heart that pulses rich red blood, no man of real human sympathies can be thrown in direct contact with an unfortunate brother in his hour of distress without responding to those noble instincts which centuries of Christian teachings have implanted in his breast.[26]

All this being said, suffering is not a good reason to pardon, *as long as the suffering is no more than deserved*. American prison conditions, it must be said, are a disgrace to humane people. Fair sentences should take into account the fact that jail is more than being behind bars; jail is being frightened, being at the mercy of guards and other prisoners, sleeping next to the toilet, sometimes being raped. But if these factors are taken into account, and offenders are serving the sentences they deserve (all things considered), then pardon is not justified by pity. As Governor John Jay said, "The power of pardoning is committed by the constitution to prudence and discretion, and not to the wishes or feelings of the governor."[27] Just as the so-called retributive emotions—vengefulness, hate, resentment—may motivate, but cannot justify punishment, emotions may motivate a pardon decision, but they cannot directly justify one.

It is important to notice, however, that an emotion like pity might *indirectly* justify a pardon. A Governor might pity an offender of subnormal intelligence who was lured into crime by a sly cellmate or a drug offender who was crippled by war-induced drug dependency and grant a pardon. But it is the factors that *caused* the pity, rather than the pity itself, that justify the pardon.

But what about the suffering of the offender's family? This factor has been cited as a justification for pardon, even by a good retributivist like Smart.[28] There is no doubt that punishing one person hurts other people. The petitions for amnesty for Vietnam draft evaders are a chronicle of other peoples' pain:

> When home on leave, applicant discovered that his wife was mentally ill and unable to care for their child. His parents were also having serious emotional problems. Applicant tried again to arrange a transfer but was told he would have to return to Vietnam and iron out the problem there.[29]

> Applicant commenced his absence from a leave status because of his father's failing health and his mother's poor economic prospects. . . . While applicant was AWOL, his father died of a stroke, leaving his mother with a pension of $22 a month. She was a polio victim and was unable to work.[30]

What will happen to these people while their sons or husbands are in jail? Does the obligation not to cause undeserved suffering in the families override the obligation to give the offenders the punishment they deserve?

Many of those who held a pardoning power have argued that it does. President William McKinley: "On his own account I do not regard him entitled to the least clemency. He has, however, a wife and eight children who are in a destitute condition."[31] President Grover Cleveland: "The granting of a pardon in this case will bring comfort to a wife and daughter whose love and devotion have never flagged, and whose affection for a husband and father remains unshaken."[32] Governor James Goodrich of Illinois: "Is society better off to let this woman struggle on with impossible conditions, the family to be broken up . . . ? The mere statement of the situation brings the answer. . . . Just so long as the law for the protection of society continues to deprive the innocent family of its sole support, with no provision for their care, just that long will these cases appeal with great and convincing force to the conscience of the executive, unless he is indifferent to human distress."[33] Even President Ford cited the very real suffering of Mrs. Nixon and the President's daughters as a reason for pardoning Nixon. (This litany of women recalls to mind the magical benefits to be gained by the debasement of medieval women, who reportedly could secure a pardon for a husband by running naked nine times around the marketplace.)

The state has always felt that it should take care of its women. This is indeed a noble sentiment in a society that makes it so hard for a woman with children to support herself. But, as Barnett pointed out sixty years ago, "the sin that society commits" should not be "considered some excuse for transforming the pardoning power into a means of poor relief."[34] Surely society should provide a safety net so that offenders' dependents do not go hungry. But pardoning the offender is an unjust way to accomplish what could be achieved by means consistent with just deserts.

Pardon on Recommendation of the Judge, Jury, or District Attorney

One of the first things the U.S. Pardon Attorney wants to know about an application for clemency is the recommendation of the judges and attorneys who decided the case in the first place. Between 1885 and 1931 (the only years for which such data are available), more federal pardons were granted because they were "recommended by the United States Attorney and Judge" than for any other single reason.[35] In the absence of any other good reason for pardoning, this seems more to be explained by the evasion of responsibility than by any reasons related to justice.

A pardon is a means for overriding the decision of a judge or jury when an offender does not really deserve the sentence imposed by legal justice. A pardon decision can take into account factors that have a moral significance, though they may lack a legal one: factors related to the whole life of the offender, his motives, and his circumstances. These facts are often not available to the judge. In some cases, a pardon decision takes into account factors that arose since the case was tried—the dying confession of the real murderer or the offender's mortal illness. In order for a judge to recommend clemency, the judge must admit either that the original decision was unjust or that the decision had unjust consequences. While some judges do recommend mercy, they have no obligation, and precious little motivation, to do so.

It is even more of a mystery why the prosecuting attorney's recommendation would justify pardon. Humbert, in his classic study of pardon,[36] pointed out the unlikelihood that a prosecuting attorney would be a font of justice:

> The United States Attorneys who frequently reach their offices because of political preferment, are often fired with a zeal to make a record by numerous convictions in order to secure further promotion. Their ardor may bring about a great number of convictions, some of which were unwarranted. But will these men be willing, afterwards, to recommend clemency in the cases in which overzealousness brought about a wrongful conviction or too severe a sentence?[37]

None of these reasons for pardon are good reasons, from a retributivist point of view, because none of them are essentially connected with what the offender deserves—and what the offender deserves is the only consideration that has the moral credentials to justify granting or withholding a pardon.

Pardon for Collected Outrageous Reasons

There remains to be considered a limited number of reasons that are sometimes cited for pardoning that are so outrageous they they would not need to be mentioned, except that they emphasize (by way of contrast) the virtues of the limits to pardoning set by a retributive theory.

The first of these so-called justifications is that the offender is a woman. Statistical studies showing whether women are pardoned in greater proportion than men have not been conducted,[38] so governors must be taken at their words, when they cite gender as a justification for pardons they grant. Governor Oswald West, for example, said of one case, "When I saw that woman in the penitentiary, . . . it made me sick, and so I turned her loose." Attorney General Dougherty recommended that a woman's death penalty be commuted, "for the sole reason that the applicant was a woman and in order to avoid the spectacle of a woman being executed."[39] Such discrimination is universally disparaged.[40] One legal scholar claimed, however, that such discrimination in favor of women need not be based on sex alone; it could be based instead on the "avowed principle that generally women do not need the same deterrent influences that are necessary to keep their brothers from going astray."[41] The retributivist, not being concerned with deterrence, need not reply.

A second outrageous reason that has been offered in support of pardon is that the offender comes from a respectable family. A pardon attorney would tremble to make such an argument publicly today, but a century ago it was assumed that well-educated, well-brought-up people were less blameworthy than less advantaged citizens, rather than vice versa.

The nineteenth century yields other reasons disconnected from desert, some of them as quaint as the iron plow: friendless condition, to permit deportation, solicitation of friends, child about to be born, to enable farmer to save crops,[42] saving of expense, prisoner bad influence on fellow prisoners,[43] etc.

This list of bad reasons for pardoning is a short one, while the human capacity for rationalizing is without visible limit. So perhaps I should add one more item to the list of improper reasons for pardoning: that is, pardon granted for any reason other than that punishment is not deserved.

So, it is possible to come up with a list of general types of pardons that cannot be defended by retributivist principles. They are as follows:

1. Pardons to promote the public welfare.
2. Pardons to promote the private welfare (of the pardoner).
3. Pardons to reward past actions.
4. Pardons for pity's sake.
5. Pardons on the recommendation of the judge, jury, or attorney.
6. Pardons based on sex or family status.

7. Pardons granted for any reason other than that punishment is not deserved.

This leaves a wide variety of types of cases in which pardons are justified by retributivist principles. Dr. Mudd deserved to be handsomely rewarded for his heroic fight against yellow fever, but not by being pardoned. He did deserve to be pardoned, however, because there is real doubt that he ever was a conspirator in Lincoln's assassination.

18

How Presidential Pardoning Practices Should Be Changed

The questions asked so far have been hypothetical: *If* retributivist principles of justice were adopted, what would pardoning practices be like? What would 'pardon' mean? When could pardons be granted? The answers have sketched a version of pardon as a duty of justice: Pardons would be granted only to people who deserved to be pardoned, for no reasons other than that they deserved to be pardoned and that it is just to give people what they deserve.

I propose, at this point, to go beyond hypothetical considerations to make some specific recommendations for reform. There are very good reasons for adopting a primarily retributivist viewpoint on matters of pardoning and for reshaping pardoning practices to reflect those principles of justice. The first reason has to do with the need to avoid incoherence and the resulting image of arbitrariness in a system of justice. Some measure of coherence was restored to the practice of *punishing* by the widespread adoption of practices justified by retributive principles. It would be well to do the same for the practice of *pardoning*. It makes no sense that a person would be put in jail for one set of reasons and then released from jail for a different set of reasons. The incoherence makes the pardoning power look unjust.

And, in fact, it *is* unjust—a second reason why the retributivist viewpoint should be adopted. A system that allows pardons to be granted for no reason at all, or for reasons having no moral legitimacy, is immoral. It licenses the executive to use citizens as means for the ends of the state, without their consent and against their own interests (as when a pardon compels testimony or returns a soldier to his military unit). It creates undeserved inequalities between offenders (as when one offender watches from behind bars as his cellmate, who is no less an offender than he, is freed). It issues an open invitation to corruption or the selfish use of power (as when a pardon prevents a criminal investigation that is potentially damaging to a powerful person[1] or political party). It disrupts and sometimes thwarts the criminal justice system's effort to give offenders what they deserve (as when an executive pardons a crony who has been fairly tried and sentenced).

Yet pardoning could be an instrument of the highest justice—a way to assist, rather than undercut, the pursuit of justice.

It is with this end in mind that I make the following recommendations for change. The recommendations are of three sorts: The first set has to do with conceptual changes, changes in ways of thinking about pardons. The second set is for specific changes in the U.S. President's pardoning power. Consistent with pardon's traditional role as a route to reform, the last set is for legal changes that are suggested by the way the presidential pardoning power is used today. The conclusion drawn is that pardon has an expanded, and much revised, role to play in America as the twentieth century draws to a close.

Recommendations for Conceptual Changes

It is important to lay to rest the hoary and destructive metaphor that represents a pardon as a shining gift from on high and a pardoning executive as a mortal god. It may have been like that when King James I held staff meetings with God, but it is like that no longer. And when the idea of pardon as a free gift goes, along with it must go the idea that pardons need no justification, being granted out of the goodness of a monarch's heart.

A pardon should be granted when justice requires it, usually to prevent the punishment of people who are innocent—if not innocent of a crime, then at least innocent of wrongdoing. This jettisons another hoary idea, that a pardon implies guilt. If pardons are properly used, very few guilty people will be pardoned, and they only to reduce their sentences.

Pardon Is Not a Gift

From a retributivist viewpoint, a justified pardon is one that corrects injustice rather than tempers justice with mercy. Retributivism calls for pardon when a reduction or remission of punishment is necessary to adjust a criminal's liability to his moral desert. So there is a duty to punish or a duty not to punish in each individual case. The executive who grants a pardon as if it were a Fabergé egg—to celebrate Easter, because he is feeling generous, or to please a family member—is acting unjustly. "The [pardoning] power of the Crown was designed to allow for the dispensation of mercy in cases where it was *deserved*,"[2] John Feerick noted in an article on the presidential pardoning power—not when it strikes the fancy of a monarch. Thus, a pardon is not a gift, but a duty of justice.

The divine metaphor, with deep roots in the English common law, should simply be left behind. If this seems a high-handed way to manipulate concepts, at least there is impressive precedent. When Justice Holmes redefined a pardon as an act for the public welfare, he purposefully disregarded both common law principles and the concepts held by the drafters of the Constitution, saying, "We will not go into history. . . . "[3] History seems equally irrelevant here; the concept of pardon as a gift leads to inconsistency and unjust practices, reasons enough to redefine the concept. The inconsistencies are as follows:

If a pardon *were* a gift, it would call for gratitude in a way that is embarrassing and inappropriate.[4] Prisoners who effusively thank governors for pardoning them imply by their actions that they have been given something more than they deserve; people are not accustomed to expressing gratitude to those who are merely doing their duty (unless they are aware of strong psychological or other temptations that were overcome in the fulfillment of the duty). Since the prisoner who does not deserve punishment has a moral right to be free from punishment, the kindness and generosity of the Governor are not at issue. Instead, the Governor's determination to act justly and steadfastness in performing moral obligations are the qualities praised.

A person who was wrongly convicted and punished should not be grateful for a pardon. On a retributivist view, all other cases of deserved pardon differ only in degree from the case of the innocent prisoner. The Governor who grants a deserved pardon is acting justly, no doubt beneficently, but certainly not benevolently.

There is a further problem in understanding a pardon as gift-giving,[5] a matter of generosity rather than justice. Gifts may be granted inconsistently, for unsavory reasons or for no reason at all. One gives the nicest valentine to the person one likes the most at the moment, and who can fathom the criteria of being likable? Perhaps having the whitest skin or the richest parents or the sweetest scent turns out to be the grounds for choosing recipients of gifts. There is no reason to criticize the criteria, even if such could be identified.

Likewise, if a pardon were an act of benevolence, there would be no wrongdoing in granting pardons inconsistently or on questionable grounds. "A little kindness is better than none" would be the guiding standard. But inconsistent pardons *are* wrong; they discriminate unfairly against people who deserve equal treatment. It is not that the pardons are granted *arbitrarily* that creates the wrong; the Roman practice of decimation by lot was brutal but at least fair. The objection is to pardons granted inconsistently or according to criteria unrelated to desert.

For these reasons, pardons should no longer be understood as gifts from the ruler.

Pardons Are Not Discretionary

Card raised this objection in uncompromising terms:

> It is not simply that we have a *right* to punish, which we are morally at liberty to exercise or not as we please. Given that we have exercised this right in some cases and that we could not justifiably refuse to exercise it in all cases, we are *obligated* to punish the offender the same as others whose cases are relevantly similar. . . . Doing the offender some good would, in itself, no more be sufficient to justify reducing or withholding a penalty than it would be sufficient to justify imposing a penalty in the first place.[6]

If one decides to honor the basic principle of justice that requires similar treatment of similars, then one cannot at the same time impose pardons for reasons of benevolence (that is to say, for no retributively significant reasons at all). An attempt to do so would result in a system like that described by Smart:

> If all identical cases of felony type-X are treated equally mercifully in 1966, to avoid unfair discrimination we should have to extend this mercy to all like cases in 1967, and all other things being equal, in 1968 and 1969 and so on, till we might say that we have changed our opinion of what is the appropriate and just penalty for this particular offense. ... But then if the appropriate penalty for identical cases of felony type-X was changed, to be consistent, we would have to adjust proportionally our idea of the fair penalty for cases of felony type-Y and type-W. ... After this new state of affairs had existed for some time we might again find people inclined to exercise mercy for no reason in some cases, and not in others that were the same. To remedy this sort of injustice ... [and so on.][7]

Smart observed that there is something "unsatisfactory"[8] about this way of making benevolent pardons fair. This is, I believe, an understatement.

The answer is to recognize that pardons, not being gifts, are not to be granted for the same reasons as gifts are. Santa Claus supposedly keeps a moral account book so that he can give presents only to the good little girls and boys, thus paying lip service to retributive justice. But in fact children's gifts depend far more on their parents' incomes and whims than on whether the children have been naughty or nice. Pardons are unlike gifts—they should not depend on income or whim, or on any factor other than that they are deserved.

Pardons Do Not Imply Guilt

There is a tremendous confusion in American law about whether or not a pardon implies guilt. William F. Duker tried to explain the untidiness:

> [I]n *Ex parte Garland* the Supreme Court noted that a pardon "blots out of existence the guilt" and makes the offender "as innocent as if he had never committed the offense." A half century after that decision, the Court said that acceptance of a pardon was an acknowledgement of one's guilt. Viewed together, the two statements are indeed paradoxical. By ascension from the hell of guilt to the heaven of innocence, one is compelled to admit his guilt to become an innocent man.[9]

The issue is a real one, since a criminal record spoils many opportunities and is likely to increase the sentence imposed for a second crime.

There has never been very good reason for inferring guilt from a pardon; pardons have always been granted for many different reasons, many for innocence. Back in 1939, legal scholar Weihofen suggested that a distinction be drawn between pardons granted for reasons of innocence and pardons granted for other reasons. A pardon for innocence, he said, "is an acquittal, and must be given all the effects of an acquittal."[10] A pardon granted for any other reason lets "the determination of the convict's guilt stand, and only relieves him from the legal consequences of that guilt."[11] This is an eminently sensible suggestion.

The only addition that needs to be made to Weihofen's recommendation is to point out that, if a retributivist's advice is taken, then by far the majority of pardons will be for innocence, legal or moral. Three of the retributivist pardon conditions are for innocence. Only one—clemency granted to reduce an unfairly harsh sentence—presupposes that the offender is guilty. Thus, a pardon alone does not imply guilt or innocence. If one were forced to make an inference, it would be safer to infer that the pardoned person is innocent, just as it is safer to infer that a meteorite will fall on water, the earth being three parts water to one part land.

The point is that there is nothing at all automatic about the inference from a pardon to guilt, or to innocence. A retributivist viewpoint severs whatever necessary connection there might have been.

Once the analogy between pardoning and gift-giving is given up, the way is clear to recommend a set of reforms. The suggestions are intended to make pardoning practices consistent with this revised, retributivist view of pardoning.

Recommendations for Changes in Pardoning Practices

Once the divine gift-giving model of pardoning is jettisoned, the model of President as divine giver-of-gifts must go as well. The American people seem unable to decide if they want their President to be like God or not. They want him to be surrounded by trumpets, no doubt, and free from mortal sin and weakness. They are unaccountably willing to assume that despite appearances, a President must know what he is doing, just as God works His wonders in mysterious ways. But in the matter of pardons, the President must be accountable to the people. Pardons are potentially too dangerous, too destructive of trust and justice, to be left to a President's whim.

Although today the President is basically free to do as he wishes, there are *some* rules about pardoning at the federal level. The constitutional grant of the pardoning power prevents the President from pardoning in impeachment cases. A federal rule requires that the President may not "exercise the pardoning power against the public interest,"[12] but leaves it up to the President to define what that might be. And the executive branch itself has issued a set of procedural rules to regulate the form and content of pardon applications.[13] However, only the first of these rules about pardoning really limits the President's power.

Several suggestions have been made for constitutional amendments that would provide more effective limits. The most notable of these is the suggestion Senator Walter Mondale made shortly after Nixon was pardoned. The proposal read: "No pardon granted an individual by the President under section 2 of Article II shall be effective if Congress by resolution, two-thirds of the members of each House concurring therein, disapproves the granting of the pardon within 180 days of its issuance."[14] Nothing came of that particular proposal, for which small favor I believe the nation may be grateful; there is no use in having more people pass judgment about a pardon until those people have a clearer understanding of what constitutes an adequate basis for judgment. However, the President's power remains unlimited, unchecked, and unreviewable.

I propose the following suggestions for ways in which the President's power to pardon should be restricted: First, pardons should only be granted after trial, conviction, and sentencing. Second, all pardons should be for specified crimes. Third, each pardon should be accompanied by a written set of reasons. Fourth, the acceptance requirement should be restored. And fifth, changes should be made in the pardoning provisions set down in the Code of Federal Regulations.

Pardon Only After Trial, Conviction, and Sentencing

The pardoning power is to the criminal justice system as an understudy is to an actor; both stand ready to step in to do the job if that becomes necessary. A pardon stands ready to do what the courts would do—protect people from punishment they do not deserve—

when the courts are prevented from doing so, either because of extraordinary circumstances surrounding the case or because laws can never be specific enough to resolve all cases satisfactorily. It follows that a pardon should not be granted until the courts have *tried* to achieve justice and have failed. Otherwise, the pardoning power would be a hindrance, rather than a help, to the criminal justice system.

Moreover, a pardon is granted on the basis of what the offender deserves, and this is often revealed by a court proceeding. A pardon that precedes a trial blocks the investigation into the facts and hides the very information that could serve to justify the pardon. So a pardon preceding trial is—literally—unjustified.

Finally, a pardon is often granted to adjust a sentence to better fit the crime. But until sentence is passed, no one knows what adjustment needs to be made.

For all these reasons, presidents should wait until an offender is tried, convicted, and sentenced before even considering a pardon.

It must be admitted that history does not agree with this limitation. When the Framers of the Constitution were debating article II, section 2, Luther Martin suggested that the power be limited to the period after "conviction."[15] James Wilson argued that the time for the exercise of the pardon power should be unlimited, so that presidents could use pardons to obtain the testimony of accomplices; Martin must have been convinced by Wilson's counterargument, because he withdrew his proposal and the subject did not come up again.

Some of the language in precedent cases approves of preconviction pardons. In *Ex parte Grossman*, Chief Justice William Howard Taft said,

> The Executive can reprieve or pardon all offenses after their commission, either before trial, during trial, or after trial, by individuals, or by classes, conditionally or absolutely, and this without modification or regulation by Congress.[16]

However, the Supreme Court has never addressed this issue squarely.

Despite this, limiting the pardon power to the period of time after sentencing would not be the drastic change that it might appear. In fact, presidents have been reluctant to pardon before conviction. The *Congressional Quarterly* reported that, "of the 2,314 pardons granted by Presidents Kennedy, Johnson, and Nixon, it appears that only three preceded conviction."[17] The President's own pardon regulations specify that an application will not even be considered until at least five years have passed after sentence was *served*. Thirty-two states have constitutional provisions that prohibit preconviction pardons. That the most spectacular American pardon of the century—the pardon of Richard Nixon—was a preconviction pardon gives the wrong impression of what generally happens.

It is sometimes argued that a pardon should be granted before charges are even brought if it appears that the accused will not be able to obtain a fair trial within a reasonable period of time. This would be a retributively respectable objection, if it were true. And, in fact, this was one of President Ford's arguments in defense of the Nixon pardon. Judge John Sirica, who probably would have been the judge in such a trial, was much offended by the charge. He answered that a fair trial would have been possible if Nixon wanted one, which he did not.[18]

The Pardon Should Be for Specified Crimes

If a pardon is only valid when it prevents the punishment of one who does not deserve to be punished, or when it limits the amount of punishment to what is deserved, then it can never be just to grant a blanket pardon for all offenses one "has committed or taken part in during [a given] period."[19] The problem is an epistemological one. If the President does not even know (or care to say) what crimes the offender has committed, how can he say that the offender does not deserve to be punished?

The English law recognized this problem early on. To prevent fraudulent applications for clemency, the general rule in England was that whenever the King was "not rightly appraised both of the heinousness of the crime, and also, how far the party stands convicted on the record, the pardon [was] void...."[20] The rationale for the rule was that pardons should be granted on the basis of the seriousness of the crime, and if the King does not know how serious the crime was, he cannot make a good judgment about whether or not to pardon.

The vagueness of Ford's pardon of Nixon is a case in point. Ford pardoned Nixon for all offenses that he had committed or may have committed, later admitting to a Senate Committee that he had no knowledge that Nixon had committed any crimes *at all*, although he knew of "ten possibilities."[21] If that is true, then it is an open admission that the pardon was granted for reasons not having to do with what Nixon deserved. And the pardon was, from the retributivist viewpoint, unjustified.

A pardon for 'anything at all' might occasionally be just, by some lucky conjunction of stars and planets, but there would be no way of knowing. And a blanket pardon always gives the appearance of injustice, looking more like an effort to cover up wrongdoing than an effort to match punishment with desert. To serve the interests, and the appearance, of justice, pardons should be granted only for specified crimes.

Pardons Should Be Accompanied by Reasons

Since a pardon is only to be granted or denied for good reason, any pardon and any refusal to pardon should be accompanied by a written account of the reasons for the decision. As it is now, offenders do not know why their applications are rejected. Nor do other interested people know why a pardon was approved.

Clearly, given the revised view of pardons, pardon decisions need justification, and the justification should be published along with the pardon decision, so that the reasons offered can be scrutinized. There is ample precedent for this among the practices of the state governors. Some governors explain every pardon decision they make, believing that they owe it to all parties concerned to demonstrate that the decision was not corrupt or arbitrary, but fully justified by sound arguments. Presidents should follow this lead and publish reasons for their decisions, as they did until around 1936.

Congress, with the help of other citizens, should study the subject of pardon and develop guidelines for its use.[22] What is needed is a public debate about what kinds of reasons are good reasons to pardon and a legislated list of those reasons. This should occur *before* the nation is embroiled in another major controversy about pardon.

Three beneficial results could be expected. First, the number of unjustified pardons would fall. The President would know in advance that people would be able to examine his reasoning in the light of generally accepted principles. That would raise the quality of debate from its present level, where people simply learn of a presidential pardon decision and like it or hate it. Second, an occasional pardon decision would not 'write.' A President might occasionally find, on attempting to defend a pardon, that it was not defensible by legitimate argument based on agreed-upon principles and would be persuaded to change his mind. Finally, if a President were to issue an indefensible pardon, there would be a clear basis for criticizing it and deciding whether or not to impose the constitutional sanction—impeachment. Any of these would be an improvement on the present situation.

The Acceptance Requirement Should Be Restored

One of the main objections to pardons as they are granted today is that they allow a President to use offenders as means to the ends of the state, without the offenders' consent and against their best interests. For example, in some cases pardon has been used to remove the Fifth Amendment right against self-incrimination.[23] In another example, President Nixon pardoned Jimmy Hoffa on condition that he have no more involvement with the labor movement, thus hoping to accomplish by a pardon what could not be accomplished by punishment.[24]

On a retributivist view, this is an illegitimate use of the pardoning power. The quickest way to eliminate it would be to restore the requirement that a pardon is not valid unless it is accepted by the person pardoned. Justice Marshall was right when he said in *U.S. v. Wilson*,

> A pardon . . . may be rejected by the person to whom it is tendered; and if it be rejected, we have discovered no power in a court to force it on him.[25]

In this way, offenders will be able to prevent themselves from being made into instruments of the President's perception of the public good.

Changes to the Pardoning Provisions, Code of Federal Regulations

The Code of Federal Regulations[26] prescribes the rules that are to govern the pardoning process. Two of the provisions are inconsistent with, and inimical to, a retributivist handling of pardons. Both should be changed, which should not be much of a jolt as the President breaks the rules regularly and with impunity.[27]

The first changes are needed in the eligibility requirements.

The Code requires a waiting period of from five to seven years after conviction, and sometimes after sentence is served, before an offender is even eligible for a pardon. Even then, the offender must show that he has a need for the pardon, or he has to wait another two years. No doubt this cuts down on the Pardon Attorney's work load, since some offenders will die or lose interest in applying for a pardon.

But the eligibility requirements cannot be defended in any other way. If a person does not deserve to be punished, he should be relieved from punishment *as soon as possible*. If his punishment is too severe, this needs to be determined before it is too late to prevent injustice. If he has been fully punished and wishes only to regain his civil rights or his reputation, then either he deserves to have them restored or he does not. If he does, then there is no justification for delay. If he does not, it may be appropriate to wait until he does, but there is no reason to think that will occur only after a pre-ordained period of time.

In short, people should be able to apply for pardons any time after they have been sentenced.

A second problem in the Code involves the guideline for making pardon decisions: Will the public welfare be better served by exercising the power of clemency for this particular case?

Only one change needs to be made: strike "the public welfare" and substitute "justice." For it is justice that should be served by the exercise of the power of clemency. One hopes and believes that in most cases justice serves the public welfare. But justice should be the primary goal, and it should be the only factor that guides a pardon decision.

Changes to Other Laws

The way the pardon power is used today points to some areas in which laws are apparently not adequate to ensure justice without presidential interference. This is true today, as it was in the past. In 1939, the U.S. Attorney General said,

> [T]his mobile institution [of pardon] has been able to play a far-reaching role in the development of the criminal law. We have seen that the law of insanity, of self-defense, of compulsion and the improved treat-

ment of the juvenile offender started from the practice of pardoning in cases where the strict application of the law seemed undesirable. This eminently creative function of the concept of pardon has not yet come to an end. The . . . aging concept . . . shrinks to make room for its new and growing offshoots.[28]

Anecdotal accounts of pardons written by governors, pardon attorneys, clemency boards, and presidents make it clear that, in some general areas of the law, presidents and governors are not satisfied with the job the laws are doing.

Frequent Pardons Point to Some Unjust Laws

Frequent pardons sometimes indicate areas in which laws should be reconsidered by legislators. It is interesting to note that governors' consciences are often good guides to injustice; like bird dogs, their pardons point to questionable legal practices. These are some of the lessons that pardons are teaching today:

There needs to be more progress on defining mental illness so that manifestly incompetent people are not convicted and punished.

People who are mentally incompetent for reasons of mental retardation or other sorts of feeble-mindedness need better protection from criminal prosecution.

So do teen-age felons.

Governors are particularly reluctant to enforce felony murder rules and other rules that make murderers out of all accomplices to a felony during which a murder takes place. It is a safe prediction that more and more who are caught in this net will be pardoned until the law is changed.

There need to be easier procedures for reopening a trial when new evidence of innocence comes to light, even long after the case has been closed.

The government's claim of right to kill offenders is questioned by many governors. Capital punishment, and particularly the way it is assigned, needs to be reconsidered.

Finally, the massive amnesties granted to illegal immigrants to the United States acknowledges the need for reform in immigration laws, which have operated unjustly since the nation began.

While all these functions—from protecting the mentally ill to determining the rights of immigrants—*could* be done by means of the pardon, it would clearly be better if they were done by alternative institutions. So the pardon has the paradoxical role of indi-rectly suggesting changes that would ultimately result in a reduced role for itself.

Changes in the Civil Rights of Convicted Felons

The vast majority of all presidential pardons are granted in order to aid convicted felons in regaining their civil rights and their good names. Most of these pardons are not granted until long after the offenders have served their time and reestablished themselves in the community.

Whether a criminal conviction costs offenders their civil rights depends on the laws of the state where they live. Some states take away an offender's right to vote, to serve on a jury, to serve as a witness, to hold a government job, to own a gun—and have no le-gal provision for ever giving them back. People who have com-pleted their sentences but find that they are still being punished by the loss of their civil rights have no recourse but through the par-don. And even with a pardon in hand, restoration of civil rights is not assured.[29]

The injustice of continuing to punish people who have served their time and the awkwardness of having to apply for a presidential pardon and then use it to apply to the state for the restoration of civil rights are manifest.

As a result, many states have developed alternative procedures for restoring civil rights.[30] Three states, Florida, Colorado,[31] and Washington,[32] automatically restore civil rights when the offender leaves prison, subject to some limitations. In England, the Rehabil-itation of Offenders Act guarantees that an offender will be "free of any handicap" resulting from a criminal conviction after a certain number of years of good behavior outside of prison.[33]

These are just and expedient approaches, which should be fol-lowed by all states. The pardoning power might still be used in ex-ceptional cases where a person deserves to have the record expunged earlier or more completely than the laws provide. Since most of the Pardon Attorney's time now is spent pardoning people who have already been punished, this one change in state law would result in the greatest change in the contemporary use of the presidential par-doning power.

The Future of Pardon

Every once in a while, someone will argue that the time has come to abolish the pardon system, since the legal system has finally reached a state close enough to perfection that pardon is no longer needed. Pardon has successfully been used as "a weapon to break the rigidity of law and custom and their resistance to progress and reform."[34] Legislate a last few changes and then it will be time, the argument goes, to let this newly flexible, progressive, reformed system of justice run unimpeded.

The argument cannot be accepted. Much of the progress in the legal system has been prompted by pardons. Pardons are being granted this very day that will have the cumulative effect of producing still further improvement in the legal system—perhaps some of those improvements suggested here. Pardon is indeed a "weapon" for reform, and its firepower has yet to be directed against the most basic structural economic and racial injustices. That battle has yet to be joined.

Conclusion

> The history of pardons shows a steady development. From irrational beginnings—touching of a holy object or person, decision by lot—it slowly grows into a human institution, regulated by reasons and reasoning.[1]

This is what the U.S. Attorney General said about pardons in 1939. If he was correct that the use of pardons is now to be governed by reasons, no one should underestimate what great progress that step alone represents. It means, first, that pardons are no longer to be understood as free acts of grace. It means, second, that pardons are subject to criticism, and that the criticism must be based on reasons and reasoning. These are both important developments.

If pardons are to be regulated by reasons, it is important to ask, exactly *what kinds* of reasons are to regulate pardons? This book has provided one possible answer: reasons having to do with justice, and only those. Feelings—either of pity or of generosity—are not good reasons for pardoning. Calculations of advantage to be gained—the pardoner's advantage or the offender's, advantages for society in general or for a political party in particular—are not good reasons. These are inadequate for pardoning, just as they are inadequate for punishing. They invite the President to use an offender as a pawn in a game not of the offender's own choosing. They lead to inequity between offenders. They frustrate the justice system's already flawed effort to match punishment with blameworthiness. And they create injustice by allowing an offender to leave his debt to society unpaid.

The only reasons that should regulate the human institution of pardon are those related to what each offender deserves. When the punishment prescribed by law is more severe than the offender deserves on account of his offense or particular circumstances, pardon is justified. Other kinds of reasons do not have the credentials to justify pardons.

A variety of objections might be raised to my claim that only reasons related to deserving should regulate pardon.

One might object in exasperation that my view of pardon dehumanizes the one last American institution with any heart, the only institution that can cut through harsh and unbending rules, responding to pain with a simple act of kindness, human being to human being. That is what pardon has always been, one might object, and that is what pardon should always be.

I would respond by allowing that a good President will be one who is kind and generous—and forgiving—in his private life. But the virtues of the leader of the government are far different. Here, the quality to be sought is integrity, the determination to do justice. The pardoning power is, in effect although not in organization, part of a system of justice, of which the penal system is also a part. The goal of the justice system is not to be kind or to secure happiness, but to insure that justice is done. There is nothing barbarous or inhumane about that. The simple act of kindness can, as Kant reminded us, "wreak injustice to a high degree."[2]

A second objection might be that a retributivist viewpoint oversimplifies. A decision about pardoning is complex, with layers of competing obligations—the obligation to act kindly as well as the obligation to do justice. A President makes the best decision who agonizes over all the different duties and then does his best.

My answer is that even this President must have some idea as to how to weigh conflicting obligations, or he would be either paralyzed (unable to decide) or he would be arbitrary (deciding for no reason in particular). Retributivism does not deny that there are obligations that conflict, particularly in regard to the most complicated issues of pardon. What retributivism does is insist that the primary duty of a state institution—both the institution of punishment and the institution of pardon—is to act justly.

When the author of *Doing Justice* set out to justify retributivist principles as the reasons that count the most in assigning punishments, he said,

> Deciding how much to punish is an agonizing process in which conflicting aspirations compete. The best one can do is decide which aspirations, on balance, appear to be most important—and build one's theory on them. Any such theory will necessarily oversimplify the moral dilemmas the decision-maker faces. Yet a rationale for allocation can bring a sense of priorities into practical decisions about punishment, difficult as these will always be.[3]

Like these authors, I do not make the mistake of thinking that even when the field of legitimate reasons for pardoning has been narrowed to one type, any decision about pardon will be an easy one. Retributive justice directs a difficult search for what it means to be blameworthy, what factors lessen blameworthiness, and how the legal and moral elements in the law should be related. These are questions deep and thick. This book has suggested some answers and has issued an invitation to others of greater insight to help define who should be pardoned on retributivist grounds and to help redesign the institution of pardoning to be responsive to those same considerations.

In many ways, a history of pardons is a history of the moral development of nations. Superstitious and arbitrary pardons marked societies just beginning to develop governments. The capricious use of pardons as acts of grace marked a time when power, in all its manifestations, was used arbitrarily and not always benevolently. If America reforms the institution of pardon to be responsive to considerations of justice above all others, then it will have taken a further step toward a more just nation.

Notes

Introduction

1. *Warrants of Pardon*.
2. Barnett, 1926, 507.
3. U.S. Attorney General, 1921–1941, 321.
4. "A Test for the Doomed," *Newsweek*, January 22, 1962, 22, 25.
5. Ford, pardon proclamation; quoted in Feerick, 1975, 8.
6. 59 *American Jurisprudence 2d*, Pardon and Parole, ss 10–72.
7. 59 *American Jurisprudence 2d*, Pardon and Parole, ss 6.
8. Sebba, 1977a, 222.
9. U.S. Department of Justice, 1985, 40; 1980, 35.
10. Personal communication, staff of the Office of the Pardon Attorney.

Chapter 1

1. Driver and Mills, 1952; quoted in Sebba, 1977a, 223.
2. Wolfgang, 1986, 1.
3. Gen. 4:12, 16.
4. Ps. 116:5.
5. But note the conditional pardon of Orestes in the *Eumenides* of Aeschylus.
6. U.S. Attorney General, 1929, 9; Abramowitz and Paget, 1964, 139.
7. Plutarch, *Numa*, 10; quoted in U.S. Attorney General, 1939, 13.
8. John 18:39.
9. Humbert, 1941, 9.
10. Alschuler, 1979, 11.
11. Hewitt, 1978, 18.
12. Davis, 1987, 12.
13. Davis, 1987, 12.
14. Tuchman, 1984.
15. Hewitt, 1978, 20.
16. U.S. Attorney General, 1939, 42. Webster's *New World Dictionary* says a "tun" is a cask of wine equivalent in liquid capacity to 252 gallons, which would be a high price indeed.
17. Davis, 1987, 10.
18. Sebba, 1977a, 224.
19. Barnes and Teeters, 1955, 362.

20. Heath, 1963, 10–11.
21. Rom. 13:1–7.
22. James I, 1918, 54–55.
23. Aquinas, *Summa Theologica*, XXVIII, q. 5–39, 90–91.
24. Leibniz, 1968, 270–271.
25. Butler, 1871, 118.
26. Manser, 1962, 305.

Chapter 2

1. Fielding, 1902, 121. Emphasis in the original.
2. Montesquieu, 1900, bk. VI, ch. 21.
3. Blackstone, 1847, 445–446.
4. *Ire partie*, title VII, art. 13. Translated by author. As quoted in Sebba, 1977a, 227.
5. Sebba, 1977a, 228. China is the one exception today.
6. Corwin, 1957, 181.
7. Hamilton, 1937, 481.
8. Hamilton, 1937, 482–483.
9. Hamilton, 1937, 482.
10. Quoted in U.S. Presidential Clemency Board, 1975, 356.
11. Kant, 1965a, 99–108.
12. von Hentig, 1937, III.
13. Heath, 1963, 10.
14. Murphy, 1987, 522.
15. U.S. Attorney General, 1939, 21.
16. Murphy, 1987.
17. Kant, 1965a, 100.
18. Kant, 1965a, 102.
19. Hodges, 1957, 209.
20. Murphy, 1972, 435; 1987, 521.
21. Kant, 1965a, 99–108.
22. Kant, 1965a, 97–98.
23. See, for example, Coddington, 1946, 155–178; Strong, 1969, 187.
24. Kant, 1965a, 107–108.
25. McCloskey, 1967, 91–110.
26. In this connection, see also an important article, Strong, 1969, 187–198.
27. McCloskey, 1967, 109.
28. McCloskey, 1967, 109.
29. Murphy (1972, 437) developed this argument, pointing out that even if one were to accept Kant's justification for punishment, it would not justify punishment in any society on the face of the earth.
30. Zweig, 1967, 199.
31. Kant, 1965a, 336.
32. Kant, 1965a, 337.
33. Kant, 1965a, 100.
34. This suggestion was offered by Michael Davis (personal communication).

Chapter 3

1. Bentham, 1948, 2–3.
2. Bentham, 1948, 152–153.
3. Beccaria, 1953, 17.
4. Fox, 1952, 11.
5. deJaucourt, 1772, 98–99.
6. deJaucourt, 1772, 98–99.
7. Howe, 1953, 808.
8. Wellman, 1975, 39–42.
9. Beccaria, 1953, 158–159.
10. Benn, 1958, 334.
11. Bentham, 1948, 171–172.
12. Bentham, 1948, 171–177.
13. deJaucourt, 1772, 98–100.
14. Wootton, 1963.
15. Wootton, 1963, 55.
16. Beccaria, 1953, 159–160.
17. Smart, 1968, 356.

Chapter 4

1. Honderich, 1970, 37.
2. Barker, 1951, 1979.
3. Hegel, 1952, 100.
4. Hegel, 1952, 219.
5. Honderich, 1970, 37.
6. Hegel, 1952, 99.
7. Hegel, 1952, 98.
8. Hegel, 1952, 220.
9. J. M. E. McTaggart, a noted Hegelian scholar, read Hegel differently. This should come as no surprise, considering the difficulty of the material. McTaggart interpreted Hegel as saying that the function of punishment is to reform the criminal, to make him finally and truly recognize the legitimacy of the laws governing him. The criminal has a right to be reformed, which imposes an obligation on the state to punish and thereby reform him. McTaggart wrote, "If he has done wrong, and if the punishment will cure him, he has, as Hegel expresses it, a right to his punishment. If a dentist is asked to take out an aching tooth, he does not refuse to do so on the ground that the patient did not deliberately cause the toothache, and that therefore it would be unjust to subject him to the pain of the extraction. And to refuse a man the chance of a moral advance—when the punishment appears to afford one—seems equally unreasonable." (McTaggart, 1896, 491)

This is a provocative suggestion about the way to ground the claim of a right to punishment. If it was Hegel's way, however, Hegel did his best (which was good indeed) to conceal the fact.

10. Like Hegel, Bernard Bosanquet was an Idealist who claimed that the proper role of punishment is to annul the crime and to show respect for the offender's will. "Punishment . . . does tend . . . to a recognition of the end by the person punished, and may so far be regarded as his own will, implied in the maintenance of a system to which he is a party, returning upon himself in the form of pain. . . . The punishment is, so to speak, his right, of which he must not be defrauded." (Bosanquet, 1923, 206–207)

11. *U.S. v. Wilson*, 32 U.S. 152 (1833).
12. *U.S. v. Wilson*, 32 U.S. 158 (1833).
13. *U.S. v. Wilson*, 32 U.S. 160 (1833).
14. U.S. Presidential Clemency Board, 1975, 176.
15. U.S. Presidential Clemency Board, 1975, 356–374.
16. *Ex parte Wells*, 59 U.S. 307 (1855).
17. U.S. Presidential Clemency Board, 1975, 361.
18. U.S. Presidential Clemency Board, 1975, 365.
19. U.S. Presidential Clemency Board, 1975, 365.
20. U.S. Presidential Clemency Board, 1975, 365.
21. Humbert, 1941, 111.
22. Humbert, 1941, 124f.

Chapter 5

1. Smithers, 1914, 66. Emphasis added.
2. Humbert, 1941, 112.
3. Bentham, 1931, 396.
4. Bentham, 1931, 430–431.
5. Alexander, 1922, 238.
6. Alexander, 1922, 241.
7. Dewey, 1950, 17.
8. Jackson, 1970, 102–103.
9. Ellwood, 1910, 536.
10. Alexander, 1922, 244–245.
11. Barnes and Teeters, 1955, 817–818.
12. Alexander, 1922, 239–240.
13. Singer, 1979, 1.
14. von Hentig, 1937, 9.
15. Menninger, 1968, 17.
16. Quoted in Weiler, 1978, 298.
17. Weihofen, 1940, 119.
18. Barnes and Teeters, 1955, front cover.
19. Bradley, 1894, 269.
20. Sharp and Otto, 1910, 341.
21. Humbert, 1941, 119.
22. U.S. Presidential Clemency Board, 1975, 374–379.
23. Abramowitz and Paget, 1964, 141.
24. As quoted in Barnett, 1926, 495.
25. Bonaparte, 1910, 605.

26. Smithers, 1914, 62–63.
27. *U.S. v. Wilson*, 32 U.S. (7 Pet.) 150 (1833).
28. *Burdick v. U.S.*, 236 U.S. 90 (1915).
29. *Biddle v. Perovich*, 274 U.S. 486.
30. *Biddle v. Perovich*, 274 U.S. 486 (1927).

Chapter 6

1. As quoted in Weiler, 1978, 304.
2. Wolfgang, 1986, 3.
3. Wolfgang, 1986, 3.
4. Zalman, 1977, 84.
5. Zalman, 1977, 88.
6. Weiler, 1978, 303.
7. *Monks v. New Jersey State Parole Board*, 58 N.J. 238 (1971).
8. Jackson, 1970, 26.
9. Lewis, 1953, 227.
10. Dershowitz, 1974, 304.
11. *In re Lynch*, 8 Cal. 3d 410 (1972).
12. *People v. Levy*, 151 Cal. App. 2d 460 (1957).
13. *People v. Levy*, 151 Cal. App. 2d 468.
14. Zimring, 1981, 105.
15. *Commonwealth v. Daniel*, 210 Pa. Super. 156, 171 n.2.
16. *State v. Chambers*, 63 N.J. 287.
17. American Friends Service Committee, 1971.
18. American Friends Service Committee, 1971.
19. American Friends Service Committee, 1971.
20. Armstrong, 1961, 471.
21. Beardsley, 1969, 36.
22. Weiler, 1978, 295.
23. Finnis, 1972, 131.
24. Gardner, 1976, 781.
25. Mabbott, 1939, 152.
26. Hart, 1960, 29.
27. See, for example, Gendin, 1970, 1; Mabbott, 1956, 289; Smith, 1965, 285; Strong, 1969, 187; Walker, 1966, 79.
28. Weiler, 1978, 316.
29. Gendin, 1970, 1.
30. Gendin, 1970, 1.
31. Finnis, 1972, 131.
32. Finnis, 1972, 135.
33. von Hirsch, 1976.
34. The Field Foundation and the New World Foundation.
35. von Hirsch, 1976, xxxix.
36. This is a Rawlsian interpretation of Kant's theory of criminal justice, very

much like that offered by Murphy (1972, 434). Murphy has subsequently backed away from this interpretation (1987, 509).

37. von Hirsch, 1976, 47.
38. von Hirsch, 1976, 47.
39. von Hirsch, 1976, 5.
40. Twentieth Century Fund, 1976.
41. Singer, 1979, XVI.

Chapter 7

1. *Ex parte Garland*, 4 Wall. 333, 380 (1867).
2. *Biddle v. Perovich*, 47 S. Ct. 665.
3. Hamilton, 1937, 481.
4. Proclamation 4311, 39 *Fed. Reg.* 32601 (1974).
5. Morris, 1926, 189.
6. Proclamation 4483, 42 *Fed. Reg.* 4393 (1977).
7. Earley, P., "Presidents Set Own Rules on Granting Clemency," *The Washington Post*, March 19, 1984, A17.
8. U.S. Department of Justice, 1985, 39; and personal communication, United States Office of the Pardon Attorney.
9. Clark, 1984, 2878.
10. U.S. Department of Justice, 1985, 39; and personal communication, United States Office of the Pardon Attorney.
11. Earley, 1984, A17.
12. Stanish, 1978, 3–7.
13. Earley, 1984, A17.
14. Zalman, 1977, 55–58.
15. *People v. Cummings*, 88 Mich. 249 (1891).
16. Sebba, 1977a, 228–237.
17. Smart, 1968, 350.
18. Roberts, 1971, 353.
19. Sebba, 1977a, 234.
20. Sebba, 1977a, 236.
21. Sebba, 1977a, 236.

Chapter 8

1. The ancient meaning for "tit" is a light stroke or blow; "tat" is an onomatopoeic variation of tap. So people who are giving tit for tat are hitting each other, in an even and unexciting fight.
2. Wasserstrom, 1964, 634–635.
3. See Murphy, 1972, 435; and Aquinas, *Summa Contra Gentiles*, lib. III, cap. 144; in Hawkins, 1944, 207.
4. See Davis, 1986, 236–267.

5. The terms 'legalistic' and 'moralistic,' come from Gardner (1976, 781), who uses them for a different purpose.

6. Hamilton, 1937, 482.

7. U.S. Attorney General, 1939, 299.

8. Smithers, 1914, 63–64.

9. Newman, 1958, 63.

10. Smart, 1968, 345.

11. Card, 1972, 182.

12. Smart, 1968, 353.

Chapter 9

1. Hart, 1961, 160–161.

2. Rawls, 1971.

3. Morris, 1968, 475.

4. Murphy, 1971, 166.

5. Finnis, 1972, 131.

6. Davis, 1983, 726.

7. See, for example, Murphy, 1971, 166–169; Narveson, 1974; and Rawls, 1964, 3–18.

8. Murphy, 1973a, 240.

9. See Murphy (1973a), who first raised the question.

10. Finnis, 1972, 132.

11. Narveson, 1974, 187.

12. See also Davis (1988, 7), who argued that the harm done is a private matter between criminal and victim, to be settled by civil processes.

13. Finnis, 1972, 132–133.

14. See *People v. Moran*, 123 N.Y. 254; *Commonwealth v. McDonald*, 5 Cush. 365 (Mass.); *People v. Jones*, 46 Mich. 441.

15. *Commonwealth v. Tibbetts*, 157 Mass. 519; *People v. Huff*, 399 Ill. 328; *Peckham v. U.S.*, 266 F. 2d 34.

16. *State v. Dumas*, 100 N.W. 2d 592.

17. Finnis, 1972, 132.

18. Finnis, 1972, 132.

19. Finnis, 1972, 132.

20. An explanation called for by Davis (1982, 88).

21. Hart, 1968, 234–235.

22. Davis, 1985, 38–39.

23. Davis, 1985, 38.

Chapter 10

1. Bradley, 1894, 269.

2. On this issue, see Kant, 1956, 63; Ross, 1930, 58.

3. See, for example, Mill (1939, 931): "Speaking in a general way, a person is understood to deserve good if he does right, evil if he does wrong; ... " Also see Aristotle, 1131^{a}9–^{b}24; Sidgwick, 1913, 279.

4. Leibniz, 1951, 353.

5. Rawls, 1971, 310.

6. Atkinson, 1969, 370.

7. Ross, 1930, 58.

8. Ross, 1930, 58.

9. The old-fashioned word 'wicked' needs explanation. It is not a particularly good choice of words because it sounds so much like Halloween to the modern ear. But no suitable substitute suggests itself. 'Bad man' sounds thuggish and excludes bad women. 'Blameworthy' begs the question. 'Vicious,' the literal opposite of 'virtuous,' sounds violent. So 'wicked' will have to serve as a general term of moral opprobrium, standing in for the phrase, 'having a bad character, being an immoral person.'

10. Frankena, 1963, 38–40.

11. Frankena, 1963, 38–40.

12. Rawls, 1971, 312–313.

13. Emmons, 1970, 133–134.

14. Murphy, 1973a, 217.

15. Lewis and Short, 1966, 1835.

16. See, for example, Bedau, 1977; Davis, 1983, 1985; Pincoffs, 1973; Alexander, 1980, 1983.

17. von Hirsch, 1976, 79.

18. von Hirsch, 1976, 79–80.

19. Singer, 1979, 21.

20. Singer, 1979, 22.

21. Singer, 1979, 23.

22. Singer, 1979, 39 (italics in original).

23. Gardner, 1976, 805.

24. Davis, 1983, 726–752.

25. Singer, 1979, 31.

26. von Hirsch, 1976, 89–90.

27. Ezorsky, 1972a, 365.

28. Parent, 1976, 350.

29. Lev. 24:17–20.

30. Kant, 1965a, 133.

31. Davis, 1983, 726–752.

32. Ferri, 1901, 12–13; quoted in Singer, 1979, 29–32.

33. Davis, 1983, 726.

34. See Ezorsky, 1972a, 365; Parent, 1976, 350.

Chapter 11

1. From a practical standpoint, it is regrettable that so much work should pile up on the shoulders of a weak-kneed word like "desert." Nobody even knows how

many *s*'s it should have—and it inspires terrible puns. For instance, a cartoon captioned "Just Desserts" shows a piece of pie saying "You are both right!"

2. This is the uncomfortable lesson one learns from Shirley Jackson's short story, "The Lottery" (1949). Every year the townspeople of the story's small community gather to draw lots. A stranger eagerly joins the lottery, only later to learn that the winner's prize is death. When the winners are the losers, "deserving to win" has a new meaning.

This is why Hegel's idea that the offender has a *right* to be punished seems so odd. Everybody is eager to get what they deserve, and the point of the principle of equity is to require that the eagerly sought portions are allocated fairly. So the language of deserving is closely connected with that of rights and entitlement. But, transferring this sort of desert (with all its desirable connotations) directly to discussions of punishment (where what is desired is usually *not* to get what is deserved) is risky.

3. Feinberg, 1970.

4. Barry, 1965, 107n.

5. Feinberg, 1970, 71.

6. Benn and Peters, 1959, 157–158.

7. Moberly, 1924–1925, 294.

8. Feinberg, 1970, 58.

9. Feinberg, 1970, 84.

10. Feinberg, 1970, 85–87.

11. Kleinig, 1971, 73.

12. Not everyone would choose these words. See, for example, Murphy (1972, 440), who speaks of 'eligibility' and 'desert.'

13. Raphael, 1955, 70–73.

14. This terminology is indirectly suggested by Wesley Hohfeld's classic analysis of fundamental legal concepts (Hohfeld, 1913, 16).

Chapter 12

1. Taft, 1892, 332.

2. Bentham, 1962, 37.

3. Borchard, 1970, VII.

4. Borchard, 1970.

5. *U.S. v. Kaplan*, 101 F. Supp. 7, 14 (1951).

6. Abramowitz and Paget, 1964, 160.

7. Borchard, 1970, 309–316.

8. Corvallis, Oregon, *Gazette-Times*, May 12, 1987.

9. Johnes, 1893, 385.

10. The phrase comes from Donnelly (1952, 20).

11. *Prisament v. United States*, 97 Cl. Ct. 434, 435 (1941).

12. *People ex rel. Prisament v. Brophy*, 287 N.Y. 132 (1941).

13. Sebba, 1977a, 229.

14. Donnelly, 1952, 20.

15. U.S. Attorney General, 1941, 75.

16. Humbert, 1941, 125.

17. Abramowitz and Paget, 1964, 160.

18. See Sebba, 1977a, 229; Donnelly, 1952, 20; U.S. Attorney General, 1939, v. III, 296.

19. U.S. Attorney General, 1941, 73.

20. Quoted in U.S. Attorney General, 1941, 75.

21. American Law Institute, Model Penal Code, draft no. 4, 156.

22. Ehrman, 1962, 21.

23. Ehrman, 1962, 21.

24. DiSalle, 1965, 59–64.

25. DiSalle, 1965, 62–63.

26. DiSalle, 1965, 64.

Chapter 13

1. Hamilton, 1937, 500.

2. *Thacker v. Commonwealth*, 134 Va. 767, 114 S.E. 504.

3. *Frazier v. State*, 86 S.W. 754 (1905).

4. *State v. Taylor*, 345 Mo. 325, 133 S.W. 2d 336.

5. *State v. Damms*, 100 N.W. 2d 592.

6. Dworkin and Blumenfeld, 1966, 396.

7. Blanshard, 1968, 66.

8. Dworkin and Blumenfeld, 1966, 397.

9. Blanshard, 1968, 66–67.

10. Dworkin and Blumenfeld, 1966, 404.

11. Murphy, 1972, 438–439.

12. Kant, 1965a, 45.

13. Murphy, 1972, 439.

14. Barnett, 1926, 521.

15. U.S. Attorney General, 1913, 341.

16. Barnett, 1926, 522.

17. Murphy, 1972, 436.

18. Murphy, 1972, 436–437.

19. Murphy, 1973a, 237.

20. President's Commission on Law Enforcement and the Administration of Justice, 1967, 44, 160; quoted in Murphy, 1973a, 237.

21. Murphy, 1973a, 237.

22. Murphy, 1973a, 242.

23. Kant, 1965, 312.

24. Davis, 1988, 12.

25. Sterba, 1984, 43; quoted in Davis, 1988, 13–14.

26. Sterba, 1984, 43.

27. Schardt et al., 1973, 39.

28. Schardt et al., 1973, 38–39.

29. Schardt et al., 1973, 40.

30. Palenberg, 1983, 369.

31. Card, 1972, 203.
32. Hart, 1968, 14.
33. *Commonwealth v. Mixer*, 207 Mass. 141, 93 N.E. 249 (1910).
34. Davis, 1987, 1363.
35. Davis, 1987, 1382–1393.
36. Davis, 1987, 1385.
37. Hall, 1957, 32.
38. *Commonwealth v. Mash*, 48 Mass. (7 Met.) 472 (1844).
39. See, generally, Wootton, 1963.
40. Blanshard, 1968, 68.
41. Frankena, 1963, 59–60.

Chapter 14

1. *Time*, May 25, 1987, 66.
2. *People v. Schmidt*, 216 N.Y. 324 (1915).
3. MacLagan, 1939, 285.
4. Quoted in Barger, 1974, 21.
5. U.S. Presidential Clemency Board, 1975, 319.
6. Barnett, 1926, 511.
7. *The Oregonian*, November 10, 1925; quoted in Barnett, 1926, 511.
8. *People v. Schmidt*, 216 N.Y. 324 (1915).
9. Blanshard, 1968, 76.
10. Franklin, 1968, 183.
11. Franklin, 1968, 184.
12. Ross, 1960, 165.
13. Ross, 1930, 165.
14. Franklin, 1968, 185–186.
15. The distinction comes from Greenawalt, 1976, 9.
16. Greenawalt, 1976, 9.
17. Franklin, 1968, 185–186.
18. *Parade Magazine*, May 3, 1987, 8.
19. *Congressional Globe*, 1871–1882, part 1, 701 (1872); quoted in Barnett, 1926, 511.
20. Barnett, 1926, 510.
21. U.S. Presidential Clemency Board, 1975, 318.
22. U.S. Presidential Clemency Board, 1975, 319.
23. U.S. Attorney General, 1939, 66.
24. U.S. Attorney General, 1939, 68.
25. Kant, 1965a, 106.
26. Kant, 1965a, 106.
27. Kant, 1965, 312.
28. Smart, 1968, 346.
29. Murphy, 1972, 441.

Chapter 15

1. Ford, "Someone Must Write, the End," *Newsweek*, (September 16, 1974), 22.
2. Ford, pardon proclamation; quoted in Feerick, 1975, 8.
3. Baker, 1974, 13.
4. Morris, 1926, 189.
5. Ted Gup, "For Seekers of Forgiveness at Lofty Levels, a Presidential Pardon," *The Washington Post*, April 17, 1987, A21.
6. Smart, 1968, 348.
7. Smart, 1968, 348.
8. Ezorsky, 1972, XXVI.
9. See Card, 1972, 201.
10. Parent, 1976, 350.
11. Parent, 1976, 353.
12. Card, 1972, 199.
13. Card, 1972, 199.
14. U.S. Attorney General, 1939, 85.
15. U.S. Attorney General, 1939, 85.
16. U.S. Presidential Clemency Board, 1975, 308.
17. U.S. Presidential Clemency Board, 1975, 127.
18. U.S. Presidential Clemency Board, 1975, 324.
19. Parent, 1976, 351–352.
20. Parent, 1976, 352.
21. Mow, 1975, 4.
22. Governor Hill of New York, quoted in Barnett, 1926, 518.
23. Humbert, 1941, 125.
24. H. B. Tedrow, State Board of Pardons of Colorado; quoted in Barnett, 1926, 505.
25. Humbert, 1941, 125.
26. Morris, 1926, 189.
27. Lempert, 1981, 1177. Alexander (1983, 233) replied to Lempert's argument.
28. DiSalle and Blochman, 1965, 50.
29. For example, Gov. Peabody of Massachusetts commuted all the death sentences handed down during his administration, as did Gov. Holmes of Oregon and Gov. Cruce of Oklahoma (Abramowitz and Paget, 1964, 176).
30. Personal communication, staff members of the Office of the Pardon Attorney.
31. *Boston Globe*, November 11, 1987.

Chapter 16

1. The reference is to Kant (1965a, 100): "The law concerning punishment is a categorical imperative, and woe to him who rummages around in the winding paths of a theory of happiness looking for some advantage to be gained by releasing the criminal from punishment. . . . "
2. Rashdall, 1900, 199. The order of the sentences is reversed.

3. Blanshard, 1968, 78.

4. Willoughby, 1910, 363–364.

5. Smart, 1968, 359.

6. Coke, 1817, 233.

7. The subject may be returning to its roots. Three recent articles debate whether God can forgive. Horsbrugh (1974, 281) argued that "a perfectly forgiving person has no occasion to forgive since he is animated by such a forgiving spirit that no conceivable injury can destroy his good will in the first place." Minas (1975, 138) agreed, arguing that God's nature makes it logically impossible that the conditions necessary for forgiveness be met—a position for which Lewis (1980, 236) takes her severely to task.

8. Horsbrugh, 1974, 281.

9. For a full development of this view, see Murphy, 1982, 503.

10. Downie, 1965, 128.

11. Flew, 1954, 293–294.

12. Downie, 1965, 130–131.

13. O'Shaughnessy (1967, 336) pointed out that the position that identifies forgiveness with the remission of punishment may take either of two forms. The first asserts that punishment remission is a necessary condition of forgiveness. The second asserts that punishment remission is a sufficient condition of forgiveness. The difference is that "the first rules out as impossible or illegitimate a statement like 'I . . . won't punish you although I can't forgive you,' while the second position rules out a statement like, 'I forgive you, but just the same I shall have to punish you.' "

14. For a lengthy and compelling argument in support of this position, see O'Shaughnessy (1967, 336–352).

15. Downie, 1965, 132.

16. Murphy, 1982, 506.

17. This account owes a great deal to Murphy (1982, 503–516).

18. Ewing, 1970, 31.

19. Horsbrugh, 1974, 271.

20. Downie, 1965, 130–131.

21. The following account owes a great deal to Hill (1978, 136–137).

22. Kant, 1964.

23. Kant, 1964.

24. Kant, 1964, 88.

25. Murphy, 1982, 513.

26. Murphy, 1986, 1.

27. Examples by the author. See also Smart, 1968, 345–359.

28. Roberts, 1971, 353.

29. The cases were suggested by Roberts, 1971, 353.

30. Smart, 1968, 345.

31. Smart, 1968, 349.

32. Roberts, 1971, 352–353.

33. Card, 1972, 182–207.

34. Card, 1972, 185.

35. Card, 1972, 185.

36. Aristotle, n.d., V, 1,137.

37. Roberts, 1971, 353.

38. *U.S. v. Wilson*, 32 U.S. (7 Pet.) 150 (1833).

39. *Biddle v. Perovich*, 274 U.S. 480 (1927).
40. Downie, 1965, 132.
41. Horsbrugh, 1974, 271.
42. Ford, "Someone Must Write, The End," *Newsweek*, (September 16, 1974), 22.
43. *Burdick v. U.S.*, 211 Fed. 492 (1915).
44. *Burdick v. U.S.*, 211 Fed. 492 (1915).
45. Feerick, 1975, 43–44.
46. Downie, 1965, 131.
47. *Ex parte Garland*, 4 Wall 333 (1867).
48. See, to start, Williston, 1914–1915, 647; Stanish, 1978, 3.
49. Ewing, 1970, 31.
50. Perelman, 1963, 58.
51. Card, 1972, 192.

Chapter 17

1. Hewitt, 1978, 116–117.
2. Quoted from a pardon petition submitted to President Carter on August 4, 1977; in Hewitt, 1978, 118.
3. *Snodgrass v. State*, 67 Tex. Cr. R. 615 (1912).
4. *Ex parte Grossman*, 267 U.S. 87, 69.
5. U.S. Attorney General, 1939, 79.
6. *Burdick v. U.S.*, 236 U.S. 79 (1915).
7. *Biddle v. Perovich*, 274 U.S. 480 (1927).
8. *Biddle v. Perovich*, 274 U.S. 665.
9. U.S. Attorney General, 1939, 77–78.
10. Michael Davis suggested that it may not have been ineptitude as soldiers that kept church burglars from being pardoned, but the fact that their crime was an offense against God, not against Regensburg (personal communication).
11. Barnett, 1926, 492.
12. *Oregonian*, December 25, 1918; quoted in Barnett, 1926, 493.
13. *San Francisco Examiner*, December 25, 1925; quoted in Barnett, 1926, 493.
14. Greenawalt, 1976, 1.
15. U.S. Attorney General, 1939, 150–153.
16. Transcript of Proceedings, Court of Impeachment, 50–51; quoted in U.S. Attorney General, 1939, 151.
17. Transcript of Proceedings, Court of Impeachment, 1844–1845; quoted in U.S. Attorney General, 1939, 153.
18. Montesquieu, 1748, bk. 6, ch. 21.
19. Barnett, 1926, 507.
20. *Governor's Conference Proceedings*, 1912, 53; quoted in Barnett, 1926, 507.
21. Abramowitz and Paget, 1964, 175.
22. Abramowitz and Paget, 1964, 175.
23. Personal communication, staff members of the Office of the Pardon Attorney.
24. U.S. Presidential Clemency Board, 1975, 309.
25. Both examples from U.S. Presidential Clemency Board, 1975, 309.

26. *Oregon Journal*, January 25, 1915; quoted in Barnett, 1926, 494.

27. Barnett, 1926, 493.

28. Smart, 1968, 345.

29. U.S. Presidential Clemency Board, 1975, 304.

30. U.S. Presidential Clemency Board, 1975, 312.

31. President McKinley, quoted in U.S. Attorney General, 1897, 183.

32. President Cleveland, quoted in U.S. Attorney General, 1894, 154.

33. Goodrich, 1920, 340–341.

34. Barnett, 1926, 496.

35. Humbert, 1941, 125.

36. Humbert, 1941.

37. Humbert, 1941, 139–140.

38. During the years 1976 to 1986, men were almost twice as likely to be granted presidential pardons as women. The average number of male offenders pardoned each of those years was 0.007 percent of the average number of male offenders in jail each year during that time. The percentage for women offenders was 0.004 percent. (*Warrants of Pardon*, 1976–1986.)

39. U.S. Attorney General, 1921, 691.

40. The question of justice that has been raised (Chapter 13) about the right of a state to punish those of its members whom it treats as second class can be raised also in regard to the right of a state to punish women.

41. Barnett, 1926, 515.

42. Humbert, 1941, 125.

43. Barnett, 1926, 527.

Chapter 18

1. During the debates about pardon in the Constitutional Convention, Iredell is reported to have said, "Pardon in a Republican government was not to protect felons with powerful friends, but to protect society from ineffective laws"; quoted in Duker, 1977, 530.

2. Feerick, 1975, 7. Emphasis added.

3. *Biddle v. Perovich*, 274 U.S. 486 (1927).

4. On this issue, see Card, 1972, 182.

5. Card, 1972, 197.

6. Card, 1972, 183–184.

7. Smart, 1968, 352.

8. Smart, 1968, 353.

9. Duker, 1977, 533.

10. Weihofen, 1939a, 179.

11. Weihofen, 1939a, 179.

12. Fisher, 1923–1924, 89–91.

13. 28 Code of Federal Regulations ss 0.35–0.36 (1973) (Pardon Attorney); and 28 Code of Federal Regulations ss 1.1–1.9 (1973) (Executive Clemency).

14. S.J. Res. 240, 93d Cong. 2d Sess. (1974).

15. 3 *The Records of the Federal Convention of 1787*, at 426.

16. *Ex parte Grossman*, 267 U.S. 120 (1925).

17. 32 *Congressional Quarterly Weekly Report* 2458 (no. 37, September 14, 1974); quoted in Macgill, 1974, 71.

18. Sirica, 1979, 231–239.

19. The phrase comes from Ford's pardon proclamation: Pres. Proc. 4311, 39 *Fed. Reg.* 32601–02 (1974).

20. Duker, 1977, 532.

21. *New York Times*, October 18, 1974, at 19, col. 2.

22. Although the Constitution grants the pardoning power to the President without any authorization to the Congress to regulate the exercise of the power, the Congress could pass legislation regulating the pardon power "even without constitutional authority so long as the power of the executive to disregard the rules is not denied." (U.S. Attorney General, 1939, 116)

23. *Burdick v. U.S.*, 27 U.S. 79 (1915).

24. *Hoffa v. Saxbe*, 378 F. Supp. 1221 (1974).

25. *U.S. v. Wilson*, 7 Pet. 161.

26. 28 Code of Federal Regulations ss. 0.35–0.36, ss. 1.1–1.9.

27. The regulations require a waiting period of five to seven years after conviction (28 Code of Federal Regulations 1.1 *et seq.* [1983]). Neither the Ford pardon of Nixon nor the Reagan pardons of FBI agents W. Mark Felt and Edward S. Miller honored this requirement.

28. U.S. Attorney General, 1939, 52.

29. Cozart, 1968, 5.

30. See 59 *Am. Jur. 2d* (Pardon and Parole) 39 (1971).

31. National Center for State Courts, 1977, 5.

32. U.S. Attorney General, 1939, 272.

33. Sebba, 1977a, 232.

34. U.S. Attorney General, 1939, 53.

Conclusion

1. U.S. Attorney General, 1939, 52.

2. Kant, 1965a, 107–108.

3. von Hirsch, 1976, 59.

Bibliography

Books and Articles

Abramowitz, E. and D. Paget: 1964, 'Executive Clemency in Capital Cases,' *New York University Law Review*, XXXIX (January), 136.

Acton, H. B.: 1969, *The Philosophy of Punishment*, Macmillan, London.

Alexander, J. P.: 1922, 'Philosophy of Punishment,' *Journal of Criminal Law and Criminology*, XIII, 235.

Alexander, L.: 1980, 'The Doomsday Machine: Proportionality, Punishment, and Prevention,' *The Monist*, LXIII, 199.

Alexander, L.: 1983, 'Retribution and the Inadvertent Punishment of the Innocent,' *Law and Philosophy*, II, no. 2, 223.

Alexander, L.: 1986, 'Consent, Punishment, and Proportionality,' *Philosophy and Public Affairs*, XV, 178.

Alschuler, A.: 1978, 'Sentencing Reform and Prosecutorial Power: A Critique of Recent Proposals for "Fixed" and "Presumptive" Sentencing,' *University of Pennsylvania Law Review*, CXXVI, 550.

Alschuler, A.: 1979, 'Plea Bargaining And Its History,' *Columbia Law Review*, LXXIX, 11.

American Friends Service Committee: 1971, *The Struggle for Justice*, Hill and Wang, New York.

American Jurisprudence, 2nd ed., Jurisprudence Publishers, 1965.

Andenaes, J.: 1958, 'Choice of Punishment,' *Scandinavian Studies in Law*, II, 59.

Anscombe, G.E.M.: 1958, 'Modern Moral Philosophy,' *Philosophy*, XXXIII, 1.

Aquinas, T.: n.d., *Summa Contra Gentiles*, III, Benziger Brothers, New York.

Aquinas, T.: n.d., *Summa Theologica*, XXVIII, Benziger Brothers, New York.

Aristotle: n.d., *Nichomachean Ethics*, V., 1130, 1131, 1137.

Armstrong, K. G.: 1961, 'The Retributivist Hits Back,' *Mind*, LXX, 471.

Armstrong, K. G.: 1972, 'The Right to Punish,' *Philosophical Perspectives on Punishment*, ed. G. Ezorsky, State University of New York Press, Albany, 136.

Atkinson, M.: 1969, 'Justified and Deserved Punishments,' *Mind*, LXXVIII, 354.

Baier, K.: 1955, 'Is Punishment Retributive?' *Analysis*, XVI, 25.

Baker, R.: 1974, 'And Me Too . . . Mr. President,' *Student Lawyer*, XIX, 13.

Barger, R.: 1974, *Amnesty: What Does It Really Mean?* Committee for a Healing Repatriation, Champaign, Illinois.

Barker, E.: 1951, *Principles of Social and Political Thought*, Clarendon Press, Oxford.

Barnes, H. and N. K. Teeters: 1951, 1955, *New Horizons in Criminology*, Prentice-Hall, Englewood Cliffs, N.J.

Barnett, J. D.: 1910, 'Grounds of Pardon in the Courts,' *Yale Law Journal*, XX, 131.

Barnett, J.: 1926, 'The Grounds of Pardon,' *Journal of Criminal Law, Criminology, and Political Science*, XVII, 490.

Barry, B.: 1965, *Political Argument*, The Humanities Press, New York.

Bayne, D.: 1966, *Conscience, Obligation, and the Law*, Loyola University Press, Chicago.

Beardsley, E.: 1969, 'A Plea for Deserts,' *American Philosophical Quarterly*, VI, 36.

Beardsley, M.: 1964, 'Equality and Obedience to Law,' in *Law and Philosophy*, ed. S. Hook, New York Institute of Philosophy, New York, 35–42.

Beccaria, C. B.: 1953, *An Essay on Crimes and Punishment*, Academic Reprints, Stanford.

Bedau, H.: 1977, 'Concessions to Retribution in Punishment,' *Justice and Punishment*, ed. J. Cederblom and W. Blizek, Ballinger, Cambridge, Mass., 51–73.

Belli, M.: 1975, 'The Story of Pardons,' *Cases and Commentary*, LXXX, 26.

Benn, S. I.: 1958, 'An Approach to the Problems of Punishment,' *Philosophy*, XXXIII, 325.

Benn, S. I. and R. S. Peters: 1959, *The Principles of Political Thought*, The Free Press, New York.

Bentham, J.: 1931, *Theory of Legislation*, Harcourt, Brace, New York.

Bentham, J.: 1948, *The Principles of Morals and Legislation*, Hafner Publishing Company, Darien, Conn.

Bentham, J.: 1962, *Works*, IX, Russell and Russell, New York, 37.

Bittner, E. and A. Platt: 1966, 'The Meaning of Punishment,' *Issues in Criminology*, II, 79.

Black, H. C.: 1968, *Black's Law Dictionary*, 4th ed., West Publishing Company, St. Paul, Minn.

Blackstone, W.: 1843, *Commentaries on the Laws of England*, IV, W. E. Dean, New York.

Blanshard, B.: 1968, 'Retribution Revisited,' in *Philosophical Perspectives* on *Punishment*, eds. E. H. Madden *et al.*, Charles C. Thomas, Springfield, Ill., 59.

Bobbio, N.: 1965, 'Law and Force,' *The Monist*, XLIX, 321.

Bonaparte: 1910, 'The Pardoning Power,' *Yale Law Journal*, IXX, 603.

Borchard, E.: 1970, *Convicting the Innocent*, Da Capo, New York.

Bosanquet, B.: 1923, *The Philosophical Theory of the State*, Macmillan, London.

Boudin, L.: 1976, 'The Presidential Pardons of James R. Hoffa and Richard M. Nixon,' *University of Colorado Law Review*, XLVIII, 1.

Bradley, F. H.: 1894, 'Some Remarks on Punishment,' *International Journal of Ethics*, IV, 269.

Bradley, F. H.: 1927, *Ethical Studies*, 2nd ed., Oxford University Press, Oxford.

Brandt, R. B.: 1959, *Ethical Theory: The Problems of Normative and Critical Ethics*, Prentice-Hall, Englewood Cliffs, N.J.

Braybrooke, D.: 1956, 'Professor Stevenson, Voltaire, and the Case of Admiral Byng,' *Journal of Philosophy*, LIII, 787.

Buchanan, G. S.: 1978, 'The Nature of a Pardon Under the United States Constitution,' *Ohio State Law Journal*, XXXIX, 36.

Butler, J.: 1871, *Analogy of Religion, Natural and Revealed, to the Constitution and Course of Nature*, Harper and Brothers, New York.

Butler, J.: 1878, *The Analogy of Religion, Two Brief Dissertations and Fifteen Sermons*, George Bell and Sons, London.

Butler, S.: 1927, *Erewhon and Erewhon Revisited*, Random House, New York.

Card, C.: 1972, 'On Mercy,' *Philosophical Review*, LXXXI, 182.

Card, C.: 1973, 'Retributive Penal Liability,' *American Philosophical Quarterly Monographs*, VII, 17.

Carey, J.: 1973, 'Amnesty: An Act of Grace,' *St. Louis University Law Journal*, XVII, 501.

Cederblom, J. and W. Blizek: 1977, *Justice and Punishment*, Ballinger, Cambridge, Mass.

Charvet, J.: 1969, 'Criticism and Punishment,' *Mind*, LXXV, 573.

Clark, C. S.: 'Reagan Parsimonious in Use of Pardon Power,' *Congressional Quarterly*, November 3, 1984, 2878f.

Coddington, F.J.O.: 1946, 'Problems of Punishment,' *Proceedings of the Aristotelian Society*, XLVI, 155.

Cohen, F.: 1940, 'Moral Aspects of the Criminal Law,' *Yale Law Journal*, XLIX, 987.

Conti, V.: 1918, 'The Concept of Punishment,' *Illinois Law Review*, XIII, 234.

Corwin, E.: 1957, *The President: Office and Powers*, New York University Press, New York.

Corwin, E.: 1974, *The Constitution and What It Means Today*, Princeton University Press, Princeton, N.J.

Cowan, T.: 1949, 'A Critique of the Moralistic Conception of the Criminal Law,' *University of Pennsylvania Law Review*, XCVII, 502.

Cowlishaw, P.: 1975, 'The Conditional Presidential Pardon,' *Stanford Law Review*, XXVIII, 149.

Cozart, R.: 1959, 'Clemency Under the Federal System,' *Federal Probation* XXIII (September), 3.

Cozart, R.: 1968, 'Benefits of Executive Clemency,' *Federal Probation*, XXXII (June), 33.

Cromwell, P.: 1985, *Probation and Parole in the Criminal Justice System*, West Publishers, St. Paul, Minn.

Cutler, S.: 1960, 'Criminal Punishment—Moral and Legal Consequences,' *Catholic Lawyer*, VI, 110.

Davis, L.: 1972, 'They Deserve to Suffer,' *Analysis*, 136.

Davis, M.: 1982, 'Sentencing: Must Justice Be Even-handed?' *Law and Philosophy*, I, 77.

Davis, M.: 1983, 'How to Make the Punishment Fit the Crime,' *Ethics*, XCIII, 726.

Davis, M.: 1985, 'Just Deserts for Recidivists,' *Criminal Justice Ethics*, III–IV, 29.

Davis, M.: 1986, 'Harm and Retribution,' *Philosophy and Public Affairs*, XV (Summer), 236.

Davis, M.: 1986a, 'Why Attempts Deserve Less Punishment than Complete Crimes,' *Law and Philosophy*, V, 1.

Davis, M.: 1987, 'Strict Liability: Deserved Punishment for Faultless Conduct,' *Wayne Law Review*, XXXIII, 1363.

Davis, M.: 1988, 'Criminal Desert, Harm, and Fairness,' paper presented at Conference on Justice in Punishment, Jerusalem, Israel, March 1988.

Davis, N.: 1987, *Fiction in the Archives*, Stanford University Press, Stanford, California.

Davis, P. E.: 1966, *Moral Duty and Legal Responsibility*, Appleton-Century-Crofts, New York.

Davitt, T.: 1960, 'Criminal Responsibility and Punishment,' *Responsibility*, ed. C. Friedrich, Liberal Arts Press, New York, 143.

deJaucourt, A. F.: 1772, 'Crime, Faute, Péché, Délit, Forfait,' *l'Encyclopédie*, ed. D. Diderot, Cramer L'aine et Compagnie, Geneva, 466–470.

del Vecchio, G.: 1956, 'Divine Justice and Human Justice," *Juridical Review*, I, 147.

Denning, Lord Justice: 1953, *The Changing Law*, Stevens, London.

Dershowitz, A.: 1974, 'Indeterminate Confinement: Letting the Therapy Fit the Harm,' *University of Pennsylvania Law Review*, CXXIII, 297.

Dewey, J.: 1950, 'The End of Punishment,' *Human Nature and Conduct*, The Modern Library, New York, 17.

Dietl, P. J.: 1970, 'On Punishing Attempts,' *Mind*, LXXIX, no. 313, 130.

DiSalle, M. and L. G. Blochman: 1965, *The Power of Life or Death*, Random House, New York.

Dix, G. and M. Sharlot: 1973, *Criminal Law: Cases and Materials*, West Publishing Company, St. Paul, Minn.

Donnelly, R.: 1952, 'Unconvicting the Innocent,' *Vanderbilt Law Review*, VI, 20.

Dorris, J.: 1953, *Pardon and Amnesty Under Lincoln and Johnson*, University of North Carolina Press, Chapel Hill.

Dostoyevsky, F.: 1950, *Crime and Punishment*, Modern Library, New York.

Downie, R.: 1965, 'Forgiveness,' *Philosophical Quarterly*, XV, 128.

Doyle, J.: 1967, 'Justice and Legal Punishment,' *Philosophy*, XLII, 60.

Driver, R. and J. Mills: 1952, *The Babylonian Laws*, Clarendon Press, Oxford.

Duerrenmatt, F.: 1960, *Traps*, Alfred A. Knopf, New York.

Duff, R. A.: 1985, *Trials and Punishments*, Cambridge University Press, Cambridge.

Duker, W.: 1977, 'The President's Power to Pardon: A Constitutional History,' *William and Mary Law Review*, XVIII, 475.

Duncan-Jones, A.: 1952, *Butler's Moral Philosophy*, Penguin Books, Harmondsworth, England.

Dworkin, G. and D. Blumenfeld: 1966, 'Punishment and Intentions,' *Mind*, LXXIII, 396.

'Effect of Pardons for Innocence Under "Habitual Criminal" Statutes,' *Yale Law Review*, LI, 1942, 699.

Ehrmann, S.: 1962, 'For Whom the Chair Waits,' *Federal Probation*, XXVI (March), 14.

Ellwood, C.: 1910, 'The Classification of Criminals,' *Journal of Criminal Law*, I, no. 4, 536.

Emmons, D. C.: 1970, 'The Retributive Criterion for Justice,' *Mind*, LXXIX, 133.

England, R. W.: 1959, 'Pardon, Commutation and Their Improvement,' *The Prison Journal*, XXXIX, 23.

Evans, I.: 1956, 'Is Punishment a Crime?' *Dublin Review*, 6.

Ewing, A. C.: 1927, 'Punishment as a Moral Agency,' *Mind*, XXXVI, 292.

Ewing, A. C.: 1963, 'Armstrong on the Retributive Theory,' *Mind*, LXXII, 121.

Ewing, A. C.: 1970, *The Morality of Punishment*, Patterson-Smith, Montclair, N.J.

Ezorsky, G.: 1972, *Philosophical Perspectives on Punishment*, State University of New York Press, Albany.

Ezorsky, G.: 1972a, 'Retributive Justice,' *Canadian Journal of Philosophy*, I, 365.

Feerick, J. D.: 1975, 'The Pardoning Power of Article II of the Constitution,' *New York State Bar Journal*, XLVII, 74.

Feinberg, J.: 1960, 'On Justifying Legal Punishment,' *Responsibility*, ed. C. Friedrich, Liberal Arts Press, New York, 152.

Feinberg, J.: 1963, 'Justice and Personal Desert,' *Justice*, ed. C. J. Friedrich and J. W. Chapman, Atherton Press, New York, 82.

Feinberg, J.: 1965, 'The Expressive Function of Punishment,' *The Monist*, XLIX, 397.

Feinberg, J.: 1970, *Doing and Deserving*, Princeton University Press, Princeton.

Feinberg, J.: 1974, 'Noncomparative Justice,' *Philosophical Review*, LXXXIII, 297.

Feinberg, J.: 1984, *Harm to Others*, Oxford University Press, Oxford.

Ferri, E.: 1901, *The Positive School of Criminology*, C. H. Kerr, Chicago.

Fielding, H.: 1902, 'Enquiry Into the Cause of the Increase of Robbers,' *The Complete Works of Henry Fielding, Esq.*, ed. W. E. Henley, Croscup and Sterling Company, New York, 119, 121.

Finnis, J.: 1972, 'The Restoration of Retributivism,' *Analysis*, XXXII, 131.

Fisher, H. A.: 1923–1924, 'The Pardoning Power of the President,' *Georgetown Law Journal*, XII, 89.

Fitzgerald, P. J.: 1962, *Criminal Law and Punishment*, Clarendon Press, Oxford.

Flew, A.G.N.: 1954, "The Justification of Punishment,' *Philosophy*, XXIX, 291.

Foote, C.: 1959, 'Pardon Policy in a Modern State,' *The Prison Journal*, XXXIX, 3.

Fox, L.: 1952, *The English Prison and Borstal Systems*, Rutledge and Kegan Paul, London.

Frankena, W.: 1963, *Ethics*, Prentice-Hall, Englewood Cliffs, N.J.

Franklin, R.: 1968, *Freewill and Determinism*, Rutledge and Kegan Paul, London.

Fuller, L.: 1964, *The Morality of Law*, Yale University Press, New Haven, Conn.

Gahringer, R.: 1969, 'Punishment and Responsibility,' *Journal of Philosophy*, LXVI, 291.

Gane, C.H.W.: 1980, 'The Effect of a Pardon in Scots Law,' *Juridical Review*, v. 1979–1981, 18.

Gardiner, G.: 1958, 'The Purposes of Criminal Punishment,' *Modern Law Review*, XXI, 117.

Gardener, G.: 1953, '*Bailey v. Richardson* and the Constitution of the United States,' *Boston University Law Review*, XXXIII, 176.

Gardner, M. R.: 1976, 'Renaissance of Retribution—An Examination of *Doing Justice*,' *Wisconsin Law Review*, v. 1976, 781.

Gendin, S.: 1967, 'The Meaning of Punishment,' *Philosophy and Phenomenological Research*, XXVIII, 235.

Gendin, S.: 1970, 'A Plausible Theory of Retribution,' *Journal Of Value Inquiry*, V, 1.

Gerber, R.: 1972, *Contemporary Punishment*, University of Notre Dame Press, Notre Dame, Ind.

Gerber, R. J. and P. D. McAnany: 1967, 'Punishment: Current Survey of Philosophy and Law,' *St. Louis University Law Journal*, XI, 491.

Gibson, S. E.: 1975, 'Constitutional Law—Presidential Pardons and the Common Law,' *North Carolina Law Review*, LIII, 785.

Gillin, J.: 1952, 'Executive Clemency in Wisconsin,' *Journal of Criminal Law*, XLII, 755.

Gingell, J.: 1974, 'Forgiveness and Power,' *Analysis*, XXXIV, 180.

Glover, M. R.: 1939, 'Mr. Mabbott on Punishment,' *Mind*, XLVIII, 498.

Golding, M.: 1977, 'Criminal Sentencing: Some Philosophical Considerations,' *Justice and Punishment*, ed. J. Cederblom and W. Blizek, Ballinger, Cambridge, Mass., 89.

Goodrich, J. P.: 1920, 'The Use and Abuse of the Power to Pardon,' *Journal of Criminal Law*, XI, 334.

Gorry, J.: 1964, 'Executive and Judicial Banishment Compared,' *Washington and Lee Law Review*, XXI, 285.

Green, T. H.: 1910, 'State's Right to Punish,' *Journal of Criminal Law and Criminology*, I, 19.

Greenawalt, K.: 1976, 'Vietnam Amnesty: Problems of Justice and Line-Drawing,' *Georgia Law Review*, XI, 1.

Gross, H.: 1979, *A Theory of Criminal Justice*, Oxford University Press, New York.

Grotius, H.: 1901, *The Rights of War and Peace*, M. Walter Dunne, New York.

Grupp, S.: 1963, 'Some Historical Aspects of the Pardon in England,' *American Journal of Legal History*, LI, 53.

Grupp, S., ed.: 1971, *Theories of Punishment*, Indiana University Press, Bloomington, Ind.

'Habeas Corpus and the Pardon Process,' *Stanford Law Review*, XI (July, 1959), 769.

Hadden, T. B.: 1965, 'A Plea for Punishment,' *Cambridge Law Journal*, v. 1965, 117.

Haines, N.: 1955, 'Responsibility and Accountability,' *Philosophy*, XXX, 141.

Hall, J.: 1957, 'Ignorance and Mistake in Criminal Law,' *Indiana Law Journal*, XXXIII, 1.

Hall, J.: 1960, *General Principles of Criminal Law*, Bobbs-Merrill, New York.

Hall, J.: 1971, 'The Purposes of a System for the Administration of Criminal

Justice,' *Theories of Punishment*, ed. S. Grupp, Indiana University Press, Bloomington, Ind., 379.

Hamilton, A.: 1937, 'The Command of the Military and Naval Forces, and the Pardoning Power of the Executive,' *The Federalist*, Random House, New York, 481.

Hare, P.: 1968, 'Should We Concede Anything to the Retributivists?' *Philosophical Perspectives on Punishment*, eds. E. H. Madden *et al.*, Charles C. Thomas, Springfield, Ill., 82.

Hart, H.L.A.: 1958, 'Legal Responsibility and Excuses,' *Freedom and Determinism*, ed. S. Hook, New York University Press, New York, 81.

Hart, H.L.A.: 1960, 'Prolegomenon to the Principles of Punishment,' *Proceedings of the Aristotelian Society*, LX, 1.

Hart, H.L.A.: 1961, *The Concept of Law*, Clarendon Press, Oxford.

Hart, H.L.A.: 1962, *Punishment and the Elimination of Responsibility*, Athlone Press, London.

Hart, H.L.A.: 1968, *Punishment and Responsibility*, Clarendon Press, Oxford.

Hart, H. M.: 1958, 'The Aims of the Criminal Law,' *Law and Contemporary Problems*, XXIII, 401.

Harword, D.: 1976, 'The Bitter Pill of Punishment,' *Journal of Value Inquiry*, CI, 199.

Hawkins, D.J.B.: 1944, 'Punishment and Moral Responsibility,' *Modern Law Review*, VII, 205.

Heath, J.: 1963, *Eighteenth Century Penal Theory*, Oxford University Press, London.

Hegel, G.W.F.: 1952, *The Philosophy of Right*, Oxford University Press, London.

Hewitt, C.: 1978, *The Queen's Pardon*, Cassell, London.

Hill, T.: 1978, 'Kant's Anti-moralistic Strain,' *Theoria*, XLIV, 136.

Hobbes, T.: 1950, *Leviathan*, E. P. Dutton, New York.

Hodge, W. C.: 1980, 'The Prerogative of Pardon,' *New Zealand Law Journal*, v. 1980, 163.

Hodges, D. C.: 1957, 'Punishment,' *Philosophy and Phenomenological Research*, XVIII, 209.

Hoffding, H.: 1912, 'The State's Authority to Punish Crime,' *Journal of Criminal Law and Criminology*, II, 691.

Hohfeld, W.: 1913, 'Some Fundamental Legal Conceptions as Applied in Judicial Reasoning,' *Yale Law Journal*, XXIII, 16.

Homer: n.d., *The Iliad*, The Modern Library, New York.

Honderich, T.: 1970, *Punishment, Its Supposed Justifications*, Harcourt, Brace, World, New York.

Hooker, R.: 1988, *The Laws of Ecclesiastical Polity, I-IV*, G. Routledge and Sons, London.

Horsbrugh, H.: 1974, 'Forgiveness,' *Canadian Journal of Philosophy*, IV, 269.

Hospers, J.: 1961, *Human Conduct*, Harcourt, Brace, World, New York.

Hospers, J.: 1977, 'Punishment, Protection, and Retaliation,' *Justice and Punishment*, eds. J. Cederblom and W. Blizek, Ballinger, Cambridge, Mass.

Howe, M.D.: 1953, *Holmes-Laski Letters*, Harvard University Press, Cambridge, Mass.

Humbert, W. H.: 1941, *The Pardoning Power of the President*, The American Council on Public Affairs, Washington, D.C.

Jackson, G.: 1970, *Soledad Brother, the Prison Letters of George Jackson*, Bantam Books, New York.

Jackson, S.: 1949, *The Lottery*, Farrar, Straus, New York.

James I: 1918, 'The True Law of Free Monarchies,' *The Political Works of James I*, ed. C. H. McIlwain, Harvard University Press, Cambridge, 53–70.

Jensen, C.: 1922, *The Pardoning Power in the American States*, University of Chicago Press, Chicago.

Jensen, C.: 1959, 'Pardon,' *Encyclopedia of the Social Sciences*, XI, Macmillan, New York, 570–572.

Johnes, E.: 1893, 'The Pardoning Power from a Philosophical Standpoint,' *Albany Law Journal*, XLVII, 385.

Jones, D. and D. Raish: 1972, 'American Deserters and Draft Evaders: Exile, Punishment, or Amnesty,' *Harvard International Law Journal*, XIII, 88.

Joyner, C. C.: 1979, 'Rethinking the President's Power of Executive Pardons,' *Federal Probation*, XLIII, 16.

Kant, I.: 1887, *Philosophy of Law*, T. T. Clark, Edinburgh.

Kant, I.: 1956, *Critique of Practical Reason*, Bobbs-Merrill, New York.

Kant, I.: 1964, *Groundwork of the Metaphysic of Morals*, Harper and Row, New York.

Kant, I.: 1965, *Critique of Pure Reason*, St. Martin's Press, New York.

Kant, I.: 1965a, *The Metaphysical Elements of Justice*, Bobbs-Merrill, New York.

Kaufman, A.: 1959, 'Anthony Quinton on Punishment,' *Analysis*, XX, 10.

Kellogg: 1977, 'From Retribution to "Desert": The Evolution of Criminal Punishment,' *Criminology*, XV, 179.

Kelsen, H.: 1957, *What is Justice?*, University of California Press, Berkeley.

Kleinig, J.: 1969, 'Mercy and Justice,' *Philosophy*, XLIV, 341.

Kleinig, J.: 1971, 'The Concept of Desert,' *American Philosophical Quarterly*, VIII, 71.

Kuhlman, A. F.: 1929, *A Guide to Material on Crime and Criminal Justice*, H. W. Wilson, New York.

Kushner, L. S.: 1981, *When Bad Things Happen to Good People*, Schocken Books, New York.

Kutner, L.: 1975, 'A Legal Note on the Nixon Pardon,' *Akron Law Review*, IX, 243.

Lamont, W. D.: 1941, 'Justice: Distributive and Corrective,' *Philosophy*, XVI, 3.

Leibniz, G. W.: 1951, 'On the Ultimate Origin of Things,' *Leibniz*, ed. P. P. Wiener, Charles Scribner's Sons, New York, 353.

Leibniz, G. W.: 1968, 'Monadology,' *Discourse on Metaphysics, Correspondence with Arnauld and Monadology*, Open Court Publishing Company, LaSalle, Ill, 249.

Lempert, R.: 1981, 'Desert and Deterrence: An Assessment of the Moral Basis of the Case for Capital Punishment,' *University of Michigan Law Review*, LXXIX, 1177.

Lessnoff, M.: 1971, 'Two Justifications of Punishment,' *Philosophical Quarterly*, XXI, 141.

Lewis, C. S.: 1953, 'The Humanitarian Theory of Punishment,' *Res Judicatae*, VI, 224.

Lewis, C. T. and C. Short: 1966, *A Latin Dictionary*, Clarendon Press, Oxford.

Lewis, M.: 1980, 'On Forgiveness,' *Philosophical Quarterly*, XXX, 236.

Locke, D.: 1963, 'The Many Faces of Punishment,' *Mind*, LXXII, 570.

Locke, J.: 1947, *The Second Treatise of Civil Government*, Macmillan, New York.

Loftsgordon, D. R.: 1966, 'Present-day British Philosophers on Punishment,' *Journal of Philosophy*, LXIII, 341.

Lyons, D.: 1969, 'Is Hart's Rationale for Legal Excuses Workable?' *Dialogue*, VIII, 496.

Maas, P.: 1983, *Marie, A True Story*, Random House, New York.

Mabbott, J. D.: 1939, 'Punishment,' *Mind*, XLVIII, 152.

Mabbott, J. D.: 1956, 'Freewill and Punishment,' *Contemporary British Philosophy*, ed. H. D. Lewis, Macmillan, New York, 289.

MacCunn, J.: 1964, *Six Radical Thinkers*, Russell and Russell, New York.

Macgill, H. C.: 1974, 'Nixon Pardon: Limits on the Benign Prerogative,' *Connecticut Law Review*, VII, 56.

MacLagan, W. G.: 1939, 'Punishment and Retribution,' *Philosophy*, XIV, 281.

Madden, E. H., R. Handy, and M. Farber: 1968, *Philosophical Perspectives on Punishment*, Charles C. Thomas, Springfield, Ill.

Mansell v. Turner, Harvard Law Review, LXXVII, 1964, 1138.

Manser, A. R.: 1962, 'It Serves You Right,' *Philosophy*, XXXVII, 293.

Mansfield, F. C.: 1977, 'Governor and the Anarchists,' *Illinois Bar Journal*, LXV (May), 600.

Marshall, J.: 1971, 'Punishment for Intentions,' *Mind*, LXXX, 597.

Martin, R.: 1970, 'On the Logic of Justifying Legal Punishment,' *American Philosophical Quarterly*, VII, 253.

'Matter of Life and Death: Due Process Protection in Capital Clemency Proceedings,' *Yale Law Journal*, XC, 1981, 889.

McCloskey, H. J.: 1965, 'A Non-utilitarian Approach to Punishment,' *Inquiry*, VIII, 239.

McCloskey, H. J.: 1967, 'Utilitarian and Retributive Punishment,' *Journal of Philosophy*, XVI, 91.

McTaggart, J.M.E.: 1896, 'Hegel's Theory of Punishment,' *International Journal of Ethics*, VI, 482.

Menninger, K.: 1968, *The Crime of Punishment*, Viking Press, New York.

Migliore, D.: 1972–1973, 'Amnesty: An Historical Justification for its Continuing Viability,' *Journal of Family Law*, XII, no. 1, 70.

Mill, J. S.: 1865, *An Examination of Sir William Hamilton's Philosophy*, Longman, Green, Longman, Roberts, and Green, London.

Mill, J. S.: 1939, 'On Utilitarianism,' *The English Philosophers from Bacon to Mill*, ed. E. A. Burtt, Modern Library, New York.

Minas, A.: 1975, 'God and Forgiveness,' *Philosophical Quarterly*, XXV (April), 138.

Moberly, W. H.: 1924–1925, 'Some Ambiguities in the Retributivist Theory of Punishment,' *Proceedings of the Aristotelean Society*, XXV, 289.

Moberly, W.: 1968, *The Ethics of Punishment*, Archon Books, Hamden, Conn.

Montesquieu: 1900, *The Spirit of Laws*, bk. VI, ch. 21, P. F. Collier and Son, New York.

Moore, G. E.: 1903, *Principia Ethica*, Cambridge University Press, Cambridge.

Morris, E.: 1926, 'Some Phases of the Pardoning Power,' *American Bar Association Journal*, XII, 183.

Morris, H.: 1961, *Freedom and Responsibility*, Stanford University Press, Stanford, Calif.

Morris, H.: 1968, 'Persons and Punishment,' *The Monist*, LII, 475.

Mow, J.: 1975, 'Rawls on Mercy: Pardon and Amnesty,' *Journal of the West Virginia Philosophical Society*, VIII, 2.

Mundle, C.W.K.: 1954, 'Punishment and Desert,' *Philosophical Quarterly*, IV, 221.

Murphy, J.: 1969, 'Criminal Punishment and Psychiatric Fallacies,' *Law and Society Review*, IV, 111.

Murphy, J.: 1970, *Kant: The Philosophy of Right*, Macmillan, New York.

Murphy, J.: 1971, 'Three Mistakes about Retributivism,' *Analysis*, XXI, 166.

Murphy, J.: 1972, 'Kant's Theory of Criminal Punishment,' *Proceedings of the Third International Kant Congress*, ed. L. W. Beck, D. Reidel, Dordrecht, Holland, 434.

Murphy, J.: 1973, 'The Killing of the Innocent,' *The Monist*, LVII, no. 4, 527.

Murphy, J.: 1973a, 'Marxism and Retribution,' *Philosophy and Public Affairs*, II, 217.

Murphy, J., ed.: 1973b, *Punishment and Rehabilitation*, Wadsworth, Belmont, Calif.

Murphy, J.: 1979, *Retribution, Justice, and Therapy: Essays in the Philosophy of Law*, D. Reidel, Boston.

Murphy, J.: 1982, 'Forgiveness and Resentment,' *Midwest Studies in Philosophy*, VII, 503.

Murphy, J.: 1985, 'Retributivism, Moral Education, and the Liberal State,' *Criminal Justice Ethics*, IV, 3.

Murphy, J.: 1986, 'Mercy and Legal Justice,' *Social Philosophy and Policy*, IV, 1.

Murphy, J.: 1987, 'Does Kant Have a Theory of Punishment?' *Columbia Law Review*, LXXXVII, 1987, 509.

Narveson, J.: 1974, "Three *Analysis* Retributivists," *Analysis*, XXXIV, 185.

Neblett, W.: 1974, 'Forgiveness and Ideals,' *Mind*, LXXXIII, 269.

Newman, C.: 1958, *Sourcebook on Probation, Parole, and Pardons*, Charles C. Thomas, Springfield, Ill.

Nino, C.: 1986, 'Does Consent Override Proportionality?' *Philosophy and Public Affairs*, XV, 178.

North, D.: 1982, *Amnesty, Conferring Legal Status on Illegal Immigrants*, Center for Labor and Migration Studies, Washington, D.C.

Nowell-Smith, P.: 1948, 'Freewill and Moral Responsibility,' *Mind*, LVII, 54.

Orland, L.: 1973, *Justice, Punishment, Treatment*, The Free Press, New York.

O'Shaughnessy, R. J.: 1967, 'Forgiveness,' *Philosophy*, XLII, 336.

Palenberg, J. C.: 1983, 'Mass Amnesty, the East German Answer to Prison Overcrowding,' *American Journal of Criminal Law*, XI (November), 369.

Paley, W.: 1818, *The Principles of Moral and Political Philosophy*, West and Richardson, Boston.

'The Pardoning Power of the Chief Executive,' *Fordham Law Review*, VI, 1937, 255.

Parent, W. A.: 1976, 'The Whole Life View of Criminal Desert,' *Ethics*, LXXXVI, 350.

Perelman, C.: 1963, *The Idea of Justice and the Problem of Argument*, Humanities Press, New York.

Piaget, J.: 1932, *The Moral Judgment of the Child*, Harcourt, Brace, New York.

Pincoffs, E. L.: 1966, *The Rationale of Legal Punishment*, Humanities Press, New York.

Pincoffs, E. L.: 1973, 'Legal Responsibility and Moral Character,' *Wayne Law Review*, XIX, 905.

Pincoffs, E. L.: 1977, 'Are Questions of Desert Decidable?' *Justice and Punishment*, eds. J. Cederblom and W. Blizek, Ballinger, Cambridge, Mass., 75.

Plato: 1959, *Gorgias*, Clarendon Press, Oxford.

Plato: 1952, *Laws*, Harvard University Press, Cambridge, Mass.

'Posthumous Pardon,' *Christian Century*, CIII, 1986, 321.

Quinton, A.: 1954, 'On Punishment,' *Analysis*, XIV, 512.

Radzinowicz, L.: 1966, *Ideology and Crime*, Columbia University Press, New York.

Raphael, D.: 1955, *Moral Judgment*, George Allen and Unwin, London.

Rashdall, H.: 1900, 'The Ethics of Forgiveness,' *International Journal of Ethics*, X, 193.

Rashdall, H.: 1924, *Theory of Good and Evil*, Clarendon Press, Oxford, 303.

Rawls, J.: 1955, 'Two Concepts of Rules,' *The Philosophical Review*, XLIV, 3.

Rawls, J.: 1958, 'Justice as Fairness,' *Philosophical Review*, XLVII, 164.

Rawls, J.: 1963, 'The Sense of Justice,' *Philosophical Review*, LXXII, 281.

Rawls, J.: 1964, 'Legal Obligation and the Duty of Fair Play,' *Law and Philosophy*, ed. Sidney Hook, New York University Press, New York, 3.

Rawls, J.: 1971, *A Theory of Justice*, Harvard University Press, Cambridge, Mass.

Records of the Federal Convention of 1787, M. Farrand, ed., Yale University Press, New Haven, 1987, 404.

Rescher, N.: 1966, *Distributive Justice*, Bobbs-Merrill, Indianapolis, Ind.

Reston, J.: 1972, *The Amnesty of John David Herndon*, McGraw-Hill, New York.

Roberts, H.R.T.: 1971, 'Mercy,' *Philosophy*, XLVI, 352.

Rose, D.: 1968, 'Retribution and Impartiality,' *The Philosophical Quarterly*, XVIII, no. 73, 356.

Ross, W. D.: 1929, 'The Ethics of Punishment,' *Philosophy*, IV, 205.

Ross, W. D.: 1930, *The Right and the Good*, Oxford University Press, Oxford.

Ross, W. D.: 1960, *Foundations of Ethics*, Clarendon Press, Oxford.

Ross, W. D.: 1968, *Aristotle*, Methuen and Company, London.

Rothman, M. D.: 1976, 'The Pardoning Power: Historical Perspective and Case Study of New York and Connecticut,' *Columbia Journal of Law and Social Problems*, XII, 149.

Saliterman, R. A.: 1985, 'Reflections on the Presidential Clemency Power,' *Oklahoma Law Review*, XXXVIII, 257.

Samek, R.: 1966, 'Punishment: A Postscript to Two Prolegomena,' *Philosophy*, XLI, 216.

Sayre, F.: 1932, *'Mens Rea,'* *Harvard Law Review*, XLV, 974.

Schardt, A., W. Rusher, and M. O. Hatfield: 1973, *Amnesty? The Unsettled Question of Vietnam*, Sun River Press, Groton-on-Hudson, N.Y.

Schoenfeld, C.: 1966, 'In Defense of Retributivism in the Law,' *Psychoanalytic Quarterly*, XXXV, 108.

Scott, A.: 1952, 'The Pardoning Power,' *The Annals of the American Academy of Political and Social Science*, CCLXXXIV, 95.

Sebba, L.: 1977, 'The Pardoning Power: A World Survey,' *Journal of Criminal Law*, LXXVIII, 83.

Sebba, L.: 1977a, 'Clemency in Perspective,' *Essays in Honor of Israel Drapkin*, eds. S. Landau and L. Sebba, Lexington Books, Lexington, Mass., 221.

Sharp, F. C. and M. C. Otto: 1910, 'A Study of the Popular Attitude toward Retributive Punishment,' *International Journal of Ethics*, XX, 341.

Sher, George: 1988, *Desert*, Princeton University Press, Princeton, N.J.

Shuman, S.: 1964, 'Act and Omission in Criminal Law: Towards a Nonsubjective Theory,' *Journal of Legal Education*, XVII, 16.

Shuman, S.: 1971, 'Responsibility and Punishment: Why Criminal Law?' *American Journal of Jurisprudence*, XV, 25.

Sidgwick, H.: 1913, *Methods of Ethics*, Macmillan, London.

Singer, R.: 1979, *Just Deserts: Sentencing Based on Equality and Desert*, Bollinger, Cambridge, Mass.

Sirica, J.: 1979, *To Set the Record Straight*, W. W. Norton, New York.

Slote, M.: 1973, 'Desert, Consent and Justice,' *Philosophy and Public Affairs*, II, 323.

Smart, A.: 1968, 'Mercy,' *Philosophy*, XLIII, 345.

Smith, A.T.H.: 1983, 'The Prerogative of Mercy, the Power of Pardon, and Criminal Justice,' *Public Law*, v. 1983 (autumn), 398.

Smith, J.: 1965, 'Punishment: A Conceptual Map and Normative Claim,' *Ethics*, LXXV, 285.

Smithers, W. W.: 1910, 'Nature and Limits of the Pardoning Power,' *Journal of Criminal Law*, I, no. 4, 549.

Smithers, W. W.: 1914, 'The Use of the Pardoning Power,' *Annals of the American Academy of Science*, LII, 61.

Sprigge, T.L.S.: 1965, 'A Utilitarian Reply to Dr. McCloskey,' *Inquiry*, VIII, 264.

Stanish, J. R.: 1978, 'The Effect of a Presidential Pardon,' *Federal Probation*, XLII, 3.

Sterba, J.: 1976, 'Justice and the Concept of Desert,' *Personalist*, LVII, 188.

Sterba, J.: 1984, 'Is There a Rationale for Punishment?' *American Journal of Jurisprudence*, XXIX, 43.

Stillman, P.: 1974, 'Prisons and Punishment,' *Journal of Social Philosophy*, V. 11.

Stillman, P.: 1976, 'Hegel's Idea of Punishment,' *Journal of the History of Philosophy*, XIV, 169.

Stoke, H.: 1927, 'A Review of the Pardoning Power,' *Kentucky Law Journal*, XVI, no. 1, 34.

Stone, W.: 1969, 'Pardon in Virginia: An Empirical Study,' *Washington and Lee Law Review*, XXVI, 313.

Strawson, P.: 1974, *Freedom and Resentment*, Methuen and Company, London.

Strong, E.: 1969, 'Justification of Juridical Punishment,' *Ethics*, LXXIX, 187.

Sutherland, E. and D. Cressey: 1955, *Principles of Criminology*, J. B. Lippincott, New York.

Taft, W. H.: 1892, *Opinions of the Attorney General of the United States*, XX, 332.

Tappan, P.: 1970, *Crimes, Justice, and Correction*, McGraw-Hill, New York.

Tarde, G. de: 1912, *Penal Philosophy*, Little, Brown, Boston.

Thalberg, I.: 1973, 'H.L.A. Hart, Punishment, and Responsibility,' *Journal of Value Inquiry*, VII, 65.

Thompson, D.: 1966, 'Retribution and the Distribution of Punishment,' *Philosophical Quarterly*, XVI, 59.

Tuchman, B.: 1984, *The March of Folly*, Knopf, New York.

Twentieth Century Fund: 1976, *Fair and Certain Punishments*, McGraw-Hill, New York.

von Hentig, H.: 1937, *Punishment: its Origin, Purpose, and Psychology*, W. Hodge and Company, London.

von Hirsch, A.: 1976, *Doing Justice*, Hill and Wang, New York.

Walker, N.: 1966, 'Varieties of Retributivism,' *The Aims of a Penal System*, Edinburgh University Press, Edinburgh.

Walker, N.: 1973, 'Review of *Doing Justice*,' *British Journal of Criminology*, XVIII, 79.

Wasserstrom, R.: 1959–1960, 'Strict Liability in the Criminal Law,' *Stanford Law Review*, XII, 730.

Wasserstrom, R.: 1964, 'Rights, Human Rights and Racial Discrimination,' *Journal of Philosophy*, 634.

Wasserstrom, R.: 1964a, 'Why Punish the Guilty?' *Princeton University Magazine*, XX, 14.

Wasserstrom, R.: 1967, 'H.L.A. Hart and the Doctrines of *Mens Rea* and Criminal Responsibility,' *University of Chicago Law Review*, XXXV, 92.

Wasserstrom, R.: 1977, 'Some Problems with Theories of Punishment,' in *Justice and Punishment*, ed. J. Cederblom and W. Blizek, Ballinger, Cambridge, Mass., 173.

Weatherburn, D.: 1984, 'Positivist and Classical Confusion in the New Probation and Parole Act,' *Criminal Law Journal*, VIII, 217.

Weihofen, H.: 1939, 'Consolidation of Pardon and Parole: A Wrongful Approach,' *Journal of Criminal Law, Criminology, and Political Science*, XXX, 534.

Weihofen, H.: 1939a, 'The Effect of a Pardon,' *University of Pennsylvania Law Review*, CLXXVII, 179.

Weihofen, H.: 1940, 'Pardon: An Extraordinary Remedy,' *Rocky Mountain Law Review*, XII, 112.

Weihofen, H.: 1956, *The Urge to Punish*, Farrar, Straus, and Cudahy, New York.

Weihofen, H.: 1960, 'Retribution is Obsolete,' *Nomos III: Responsibility*, ed. C. J. Friedrich, Liberal Arts Press, New York, 116.

Weiler, J.: 1978, 'Why Do We Punish: The Case for Retributive Justice,' *University of British Columbia Law Review*, XII, 295.

Weiler, P.: 1974, 'The Reform of Punishment,' *Studies on Sentencing: Working Paper #3*, Law Reform Commission of Canada, Ottawa.

Weisman, N.: 1972, 'History and Discussion of Amnesty,' *Columbia Human Rights Law Review*, IV (Fall), 529.

Wellman, C.: 1975, *Morals and Ethics*, Scott, Foresman, Glenview, Ill.

Whiteley, C. H.: 1948, 'Nowell-Smith on Retribution and Responsibility,' *Mind*, LVII, 230.

Whiteley, C. H.: 1956, 'On Retribution,' *Philosophy*, XXXI, 154.

Williston, S.: 1914–1915, 'Does A Pardon Blot out Guilt?' *Harvard Law Review*, XXVIII, 647.

Willoughby, W.: 1910, 'Anglo-American Philosophies of Penal Law. II,' *American Institute of Criminal Law and Criminology*, I, 354.

Willoughby, W.: 1929, *The Constitutional Law of the United States*, Baker, Voorhis and Company, New York.

Wilson, J. Q.: 1975, *Thinking About Crime*, Basic Books, New York.

Wolfgang, M.: 1959, 'Murder, the Pardon Board and Recommendations by Judges and District Attorneys,' *Journal of Criminal Law, Criminology, and Political Science*, C, no. 4, 338.

Wolfgang, M.: 1986, 'A Return to Just Deserts,' *The Key Reporter*, LII, 1.

Wolfgarten, A. and A. N. Khan: 1986, 'Free Pardon,' *The Solicitor's Journal*, CXXX, 157.

Wootton, B.: 1959, *Social Science and Social Pathology*, Macmillan, New York.

Wootton, B.: 1963, *Crime and the Criminal Law*, Stevens, London.

Zalman, M.: 1977, 'The Rise and Fall of the Indeterminate Sentence,' *Wayne Law Review*, XXIV, 45.

Zimring, F.: 1977, *Making the Punishment Fit the Crime*, University of Chicago Press, Chicago.

Zimring, F.: 1981, 'Sentencing Reform in the States,' *Northern Illinois University Law Review*, v. II, 101.

Zweig, A.: 1967, *Kant: Philosophical Correspondence, 1759–1799*, University of Chicago Press, Chicago.

Legal Cases

Biddle v. Perovich, 274 U.S. 480 (1926).

Bird v. State, 190 A.2d 804 (1963).

Bjerkan v. U.S., 529 F.2d 125 (1975).

Burdick v. U.S., 221 Fed. 492, 236 U.S. 79 (1915).

Ex parte Chain, 49 So.2d 722 (1951).

Chapman v. Scott, 10 F.2d 156 (1925).

Commonwealth v. Daniel, 210 Pa. Super. 156, 232 A.2d 247 (1967).

Commonwealth v. Mash, 48 Mass. (7 Met.) 472 (1844).

Commonwealth v. McDonald, 5 Cush. 365 (Mass.) (1850).

Commonwealth v. Mixer, 207 Mass. 141, 93 N.E. 249 (1910).

Commonwealth v. Tibbetts, 157 Mass. 519, 32 N.E. 910 (1893).

Connecticut Board of Pardons v. Dumschat, 101 S. Ct. 2460 (1981).

Ex parte Crump, 135 P. 428 (1913).

D.P.P. v. Smith, A.C. 290 (1961).

Dean v. Skeen, 70 S.E.2d 256 (1952).

Diehl v. Rodgers, 169 Pa. St. 323 (1895).

In re Dupuy, Fed. Case No. 3, 814, 7 Fed. Cas. 506 (1869).

Frazier v. State, 86 S.W. 754 (1905).

Frazier v. Warden of Maryland Penitentiary, 109 A.2d 78 (1954).

Ex parte Garland, 71 U.S. (4 Wall) 333 (1866).

Gregg v. Georgia, 96 S. Ct. 2909 (1976).

Ex parte Grossman, 267 U.S. 87 (1925).

Hoffa v. Saxbe, 378 F. Supp. 1221 (1974).

Kelley v. State, 185 N.E. 453 (1933).

Lambert v. California, 355 U.S. 225, 78 S. Ct. 240 (1957).

In re Lynch, 8 Cal.3d 410 (1972).

Monks v. New Jersey State Parole Board, 58 N.J. 238 (1971).

Moore v. Lawrence, 15 S.E. 2d 519 (1941).

Morisette v. U.S., 342 U.S. 246 (1942).

Murphy v. Ford, 390 F. Supp. 1372 (1975).

Peckham v. U.S., 96 U.S. App. D.C. 312, 226 F.2d 34 (1955).

People v. Broncado, 188 N.Y. 150, 80 N.E. 935 (1907).

People v. Cummings, 88 Mich. 249 (1891).

People v. Huff, 339 Ill. 328, 171 N.E. 261 (1930).

People v. Jones, 46 Mich. 441, 9 N.W. 486 (1881).

People v. Levy, 151 Cal. App.2d 460 (1957).

People v. Molinas, 250 N.Y.S.2d 684 (1964).

People v. Moran, 123 N.Y. 254, 25 N.E. 412 (1890).

People v. Nowak, 55 N.E.2d 63 (1944).

People v. Schmidt, 216 N.Y. 324, 110 N.E. 945 (1915).

People v. Silverman, 181 N.Y. 235 (1905).

People ex rel. Prisament v. Brophy, 487 N.Y. 132 (1941).
Prisament v. United States, 97 Cl. Ct. 434 (1941).
Regina v. Pembliton, 12 Cox Crim. Cases 607 (1874).
Regina v. Ward, 1 Q.B. 351 (1956).
Roberts v. State, 160 N.Y. 217 (1899).
Schick v. Reed, 419 U.S. 256 (1974).
Snodgrass v. State, 67 Tex. Cr. R. 615 (1912).
State v. Chambers, 63 N.J. 287, 307 A. 2d 78 (1973).
State v. Damms, 9 Wis.2d 183, 100 N.W.2d 592 (1960).
State v. Harrison, 115 S.E. 746 (1923).
State v. Taylor, 345 Mo. 325, 133 S.W.2d 336.
Thacker v. Commonwealth, 134 Va. 767, 114 S.E. 504 (1922).
U.S. v. Athens Armory, Fed. Cas. #14, 473; 884 (1868).
U.S. v. Kaplan, 101 F. Suppl. 7 (1951).
U.S. v. Klein, 80 U.S. 128 (1869).
U.S. v. Patti, 291 F.2d 745 (1961).
U.S. v. Wilson, 32 U.S. (7 Pet.) 150 (1833).
Ex parte Wells, 18 How. 307 (1855).
Wilborn v. Saunders, 195 S.E. 723 (1938).

Public Documents

American Law Institute, *Model Penal Code*, draft no. 4, p. 156.
Code of Federal Regulations, Title 28—Judicial Administration.
Coke, E.: 1817, *The Third Part of the Institutes of the Laws of England*.
Jones, T.: 1870, *Amnesty*, Congressional Globe Office, Washington, D.C.
Massachusetts Parole Board: 1978, Decision-Making Guidelines and Procedures, Boston.
National Center for the State Courts: 1977, *Clemency: Legal Authority, Procedure, and Structure*, Williamsburg, Va.
Neal, A.: 1929, *A Summary of the Provisions of the Constitution and Statutes of the Several States Relating to Pardons*, Wisconsin Legislative Reference Library, Madison.
President's Commission on Law Enforcement and Administration of Justice: 1967, *The Challenge of Crime in a Free Society*, U.S. Gov't. Printing Office, Washington, D.C.
U.S.: 1971, *Proposed Federal Criminal Code*, sec. 302.
U.S. Attorney General: 1894–1910, 1911–1920, 1921–1941, *Reports of the Attorneys General of the United States*, Office of the Pardon Attorney, Chevy Chase, Md.
U.S. Attorney General: 1939, *Survey of Release Procedures*, v. III, U.S. Gov't. Printing Office, Washington, D.C.

U.S. Attorney General: 1942–1985, *Annual Report of the Attorney General of the United States*, U.S. Gov't. Printing Office, Washington, D.C.

U.S. Attorney General: 1960, *Survey of Release Procedures*, v. III, U.S. Gov't Printing Office, Washington, D.C.

U.S. Constitution, Article II, sec. 2.

U.S. National Archives: 1955, *Preliminary Inventory of the Office of the Pardon Attorney* (no. 37, Record Group 204), compiled by G. Kerner, Washington, D.C.

U.S. Presidential Clemency Board: 1975, *Report to the President*, U.S. Gov't. Printing Office, Washington, D.C.

The War of the Rebellion: 1907, series III, vol. V., United States Gov't. Printing Office, Washington, D.C.

Warrants of Pardon: 1942–1987, Office of the Pardon Attorney, Chevy Chase, Maryland.

Unpublished Materials

Card, C.: 1969, "Retributive Justice in Legal Punishment," unpublished Ph.D. dissertation, Department of Philosophy, Harvard University.

Loftsgordon, D. R.: 1958, "Retributive Morality and its Alternatives," unpublished Ph.D. dissertation, Department of Philosophy, Columbia University.

Moore, J. B.: 1965, "On Retributive Justifications of Punishment," unpublished Ph.D. dissertation, Department of Philosophy, Harvard University.

Narveson, J.: 1974, "Retribution, Fairness, and Utilitarianism," paper read before the Western American Philosophical Association Meeting, St. Louis, Mo.

Seligman, D.: 1966, "Justice and the Role of Retribution in Punishment," unpublished Ph.D. dissertation, Department of Philosophy, Columbia University.

Veblen, T.: 1884, "The Ethical Grounds of a Doctrine of Retribution," unpublished Ph.D. dissertation, Department of Philosophy, Yale University.

Index

DATE DUE